Environmental Guilt and Shame

Environmental Guilt and Shame

Signals of Individual and Collective Responsibility and the Need for Ritual Responses

SARAH E. FREDERICKS

OXFORD
UNIVERSITY PRESS

OXFORD
UNIVERSITY PRESS

Great Clarendon Street, Oxford, OX2 6DP,
United Kingdom

Oxford University Press is a department of the University of Oxford.
It furthers the University's objective of excellence in research, scholarship,
and education by publishing worldwide. Oxford is a registered trade mark of
Oxford University Press in the UK and in certain other countries

First Edition published in 2021

Impression: 1

Published in the United States of America by Oxford University Press
198 Madison Avenue, New York, NY 10016, United States of America

British Library Cataloguing in Publication Data
Data available

Library of Congress Control Number: 2020951908

ISBN 978–0–19–884269–9

DOI: 10.1093/oso/9780198842699.001.0001

Printed and bound by
CPI Group (UK) Ltd, Croydon, CR0 4YY

Acknowledgements

First, thank you to all of the everyday environmentalists who, when finding out what I did for a living, confessed their guilt and/or shame to me and asked for my advice. While I did not ask for such confessions or questions, continually experiencing them in random encounters, after many popular (and academic) presentations on the material that became this book, and in conversations with many friends and family members made me realize that there was something important here to study.

This book has been a long time in coming. Thanks to the organizers of the 2012 ISSRNC conference on Nature and the Popular Imagination, whose call for proposals led me to think more expansively about my research and prompted me to start down this path.

While writing I have been fortunate to been a part of several amazing writing groups. Teresa D. Golden masterfully led a writing group at the University of North Texas. Her feedback and that of others in the group were instrumental in facilitating my writing as I began this second major project. I appreciate the leadership of the University of North Texas who funded the peer mentoring program which gave financial support to our group. In Chicago, my writing group, including Angie Heo, Alizrea Doostdar, Elham Mireshghi, Emrah Yildiz, Adia Benton, Mareike Winchell, Elina Hartikainen, Andy Graan, and especially Ghenwa Hayek and Darryl Li, have provided encouragement and helpful feedback, particularly on the material about everyday environmentalists. Lea Schweitz deserves more thanks than I can give for her companionship and insights regarding writing about religion and nature.

Colleagues and students in the Department of Philosophy and Religion at UNT, especially J. Baird Callicott and Robert Figueroa, were instrumental in my background thinking for this project. My many conversations with Irena Klaver after our classes at the gym pushed me to hone my arguments and aim high. At UNT, research assistants Nick Sarratt, Gordon Akon-Yamga, Seth Thompson, and especially Sarah Conrad helped gather research for this project. Nancy Ellis's support, administrative and otherwise, was invaluable. I appreciate Trish Glazebrook's prioritization of research funding for junior faculty within our department and a 2014 UNT Scholarly and Creative Activity Grant which made such RA support possible.

The rich intellectual conversations with colleagues and students at the University of Chicago have nourished this project. Particular thanks go to Richard Rosengarten, Margaret Mitchell, Richard Miller, Kevin Hector, William Schweiker,

Laurie Zoloth, Cynthia Lindner, Karin Krause, Curtis Evans, everyone at the faculty retreat who discussed my work, the participants in the Environmental Studies workshop, Theology and Ethics workshop, and students in my Collective Guilt, Shame and Responsibility and Collective Agency and Responsibility courses. Colleagues across the university and Hyde Park, including Fredrik Albritton Jonsson, Sabina Shaik, Julia Adney Thomas, and Elizabeth Chatterjee, have also been lively conversation partners. My research assistants Amanda Gvozden, Caroline Anglim, and particularly Colin Weaver and Katrina Myers, have located sources, proofread, and served as invaluable sounding boards as I worked through the arguments and structure of this book. Sara-Jo Swiatek helpfully indexed the book. Research funds from the Divinity School as allocated by our Deans Margaret Mitchell, Richard Rosengarten, and David Nirenberg enabled this RA assistance and my participation in many valuable conferences. Thank you. Support staff at the Divinity School, particularly Nathelda McGee, Sara Bigger, and Suzanne Riggle, facilitated my research and writing.

I deeply appreciate my colleagues in the study of environmental ethics, religion, and the environment or religious studies more broadly, including Ben Almassi, David Barr, Evan Berry, Christopher Carter, Forrest Clingerman, Celia Deane-Drummond, Rachel Fredericks, Tallessyn Grenfell-Lee, Laura Hartman, Willis Jenkins, Hee Kyung Kim, Takeshi Kimura, Catherine Klancer, Derek Michaud, Cynthia Moe-Lobeda, Kevin O'Brien, Sarah Pike, Daniel Scheid, Lisa Sideris, Willa Swenson-Lengyel, Bron Taylor, Sarah McFarland Taylor, and Kyle Powys Whyte. Additionally, the development of this work was facilitated by co-panelists, moderators, audience members, and organizers of the American Academy of Religion, the International Society for the Study of Religion, Nature, and Culture, the International Society for Environmental Ethics at the 2016 Central American Philosophical Association meeting, "Vcologies: Disciplinary Approaches to Environmental Knowledge," The Institute on Religion in an Age of Science, the Midwest Faculty Seminar on the Environmental Humanities at the Franke Institute for the Humanities, the Disciples Divinity House, the Japanese Association for Religious Studies, the North Texas Philosophical Association, the Dallas Philosophers Forum, and events at Elmhurst College, Duquesne University, the Chan Buddhist Temple of Changshu and Soochow University, the Kokoro Research Center at Kyoto University, Francis Marion University, the Consortium for the Study of Religion, Ethics and Society and the Poynter Center at Indiana University, Rice University, and Gustavus Adolphus College. "Being Human in the Age of Humans: Perspectives from Religion and Ethics," a project funded by the Humanities Without Walls Consortium, and led by PI Lisa Sideris and Project Coordinators Kyle Powys Whyte, Celia Deane-Drummond, and myself, enriched my thinking about the Anthropocene in ways that shaped Chapter 5 in particular.

The hospitality and beautiful nature-focused retreat center at Holy Wisdom Monastery made it a generative place to write at two key times during this project.

Anonymous reviewers and Stephen Pattison were generous with their time as they read and commented upon this manuscript. Thanks also to Karen Raith and the rest of the team at Oxford for supporting and shepherding this project.

Thank you to the following publishers for permissions to reuse and develop previously published materials. "Online Confessions of Eco-Guilt," *Journal for the Study of Religion, Nature and Culture* 8(1) (2014): 64–84 ©Equinox Publishing Ltd [2014] was a preliminary study for the project; Chapters 2 and 5 in particular expand upon this text. "Climate Apology and Forgiveness," originally published in *Journal of the Society of Christian Ethics* 39(1), (Spring/Summer 2019): 143–59, is expanded upon in sections 4.3 and 4.4. Its arguments also inform Chapter 9. Chapters 6, 8, and 9 draw upon ideas developed in "Ritual Responses to Geoengineering" in *Calming the Storm: Theological and Ethical Reflections on Climate Engineering*, edited by Forrest Clingerman and Kevin J. O'Brien, Lexington Press, 2016, pp. 165–86, all rights reserved. Several paragraphs from my 2020 chapter "Thinking at the Margins of Sustainability in Religious Ethics: Indicators, Emotions, Rituals," published in *The Global Sustainability Challenge*, edited by Gerard Magill and James Benedict, are published with the permission of Cambridge Scholars Publishing.

Last, but certainly not least, my friends and family have given me encouragement, space, and helped me attend to other parts of my life as is certainly necessary in general and to write. I specifically thank those not mentioned above: the Denton Quilt Guild; Alex Ponette-Gonzalez; Matt Fry; Brian O'Connor; Jennifer Jensen Wallach; Kelly Wisecup; Gayatri Mehta; the Lakeside Quilting Guild; Aubrey Pech; Julie Sorrell; Susan, Larry, Ethan, and Morgan Engelhardt; Crissy, Chris, and Nathan Hutchison-Jones; Penny Bryant; Stephanie Beal; and Dave, Sylvia, and Rachel Fredericks.

Contents

List of Tables

1
Introduction

Early in *No Impact Man: The Adventures of a Guilty Liberal who Attempts to Save the Planet and the Discoveries He Makes About Himself and Our Way of Life in the Process,* Colin Beavan describes tracking his family's trash for four days to understand how to reduce their waste. Expecting to do much better than the average U.S. American, he opens his family's trash bags (except for the bag containing leaking diapers) and inventories their contents.[1]

> Surrounded by all this debris, sitting there on my hallway floor with a pen and paper in my hand, I felt like I'd just stepped on the scale and the news was much worse than I'd thought. It was not trash per se that got me. It was the throwing away of things used for less than five minutes without so much as a thought before reaching for the exact same product to use for another five minutes before throwing that away, too. The truth was that every coffee cup and every water bottle in the corner trashcan gave me a tiny micro-twinge of guilt.
>
> (Beavan 2009, 35–6)

After recognizing that his wasteful actions make him feel guilty, he deepens his reflection. He remembers his childhood relationship with his grandparents who lived through the Depression. They held values not just against wasting but also "that we ought not to take for granted what we had been blessed with". He wonders "What would my grandparents think of me now?" (Beavan 2009, 37).

> You sit down with your trash and you see your life laid before you on the floor, you see what an archeologist would see when he studies your life a thousand years from now, and you wonder: If life begets life and death begets death, does waste beget waste? If my life begets waste, what does that say about my life? Is a waste of resources a sign of a waste of life?
>
> Half an hour on the floor with my trash and I'd gotten myself stuck in an existential morass. (Beavan 2009, 43)

[1] I follow Roxanne Dunbar-Ortiz in using the term "U.S. American" to designate people who live in the United States to avoid the imperialism of the term "American" vis-à-vis our neighbors throughout the North and South American continents (Dunbar-Ortiz 2014, xiii–xiv).

Environmental Guilt and Shame: Signals of Individual and Collective Responsibility and the Need for Ritual Responses.
Sarah E. Fredericks, Oxford University Press (2021). © Sarah E. Fredericks. DOI: 10.1093/oso/9780198842699.003.0001

He tries to get out of the existential paralysis of shame by contextualizing his situation:

> [...] it is not that my family alone had turned into some sort of monstrous, garbage-making machine. It's not that I'm a marred human being who took a wrong turn, or that I've turned bad in the twenty-five years since my grandparents wielded their influence over me. It's not that I'm the lazy ingrate I thought I was. But it may be that as a member of the crew of the huge steamship that is our culture, I had acquiesced to some decisions that caused the whole boat to take a wrong turn, and possibly sink. (Beavan 2009, 40)

Here and elsewhere, Beavan alludes to the temptation to judge his character, as well as that of U.S. American society and humans in general, as completely problematic: a "lazy ingrate" or "failure" (Beavan 2009, 40, 64, 190), not just one who does some bad actions. Yet, he repeatedly asserts that both he and humanity are good, or at least neutral: "I'm an optimist about human nature. Our desires are fundamental to who we are and therefore essentially good (even if we should not necessarily allow ourselves to be ruled by them)" (25). He proclaims that "trash and pollution and greenhouse gases, therefore, are not a result of some total flaw in human nature," but rather arise from acting as if infinite resources, including those for waste processing, were available; a flaw of the system not the species (65–6). Thus, environmental degradation is not just the result of individual actions but is a problem of the whole, of a collection of humans. For Beavan, these problematic actions raise existential questions about human character and nature.

Beavan's text is far from unique in its reflection on negative judgments and emotions about the actions of individual humans, groups of humans (a family, U.S. American society, Western industrialized nations), and sometimes humanity as a species. Indeed, many authors of popular environmental texts, including memoirs, self-help books, newspaper articles, and blog and discussion board posts, indicate that concern about environmental actions and what such actions say about individual and collective character are common features of life among middle- and upper-class English-speaking people in the United States and other developed nations. The "everyday environmentalists" I study are "people who attempt to live out their environmental concern through ordinary, routine behaviors such as eating, shopping, cleaning, and transportation" (Fredericks 2014b, 65). They are not environmental professionals or radical activists or even substantially engaged in activist work. Their narratives do not, however, just reveal conceptual or theoretical problems. They often reveal that people's visions of themselves are so negative that they are unable to conceive of, let alone enact, positive change. For instance, after writing a year-long newspaper column about trying to lead a better life with respect to the environment, Leo Hickman

> received well over six hundred letters and emails from people across the world [...] they too often felt a similar feeling of inertia and guilt [shame] about not being able to lead a more ethical life. Some felt they had made a successful switch in their lives, whereas others felt that their efforts had been a miserable, even worthless failure. (Hickman 2005, 7)

Indeed, frequently people are paralyzed in the face of the horror of their participation (and/or that of their society) in environmental degradation. The familiarity of these experiences is documented in many popular texts including environmental memoirs or how-to books such as Jen Pleasants' *Bag Green Guilt: 5 Easy Steps: Turn Eco-Anxiety into Constructive Energy*; Paige Wolf's *Spit That Out! The Overly Informed Parent's Guide to Raising Children in the Age of Environmental Guilt*; Robyn Harding's *Mom, Will This Chicken Give me Man Boobs?: My Confused, Guilt-Ridden, and Stressful Struggle to Raise a Green Family*; and Fritzie von Jessen's *I Killed A Penguin: An Ecological Memoir.*

In these texts, and many others analyzed in detail in Chapters 2 and 5, we see individuals judging their *actions* as individuals and those of their society as a collective to be wrong (guilt) and judging their very *selves* and the identity of their collective as flawed (shame). While these verdicts may be catalyzed by the presence of others or actively induced by others, as when one person deliberately shames another, the agents' own judgments and ideals are the critical elements in their judgment that they are guilty, shameful, or both. Authors frequently also *feel* bad about their own and their collective's actions (guilt) and what such actions say about themselves and their society's character (shame). Their individual and collective experiences of both action-oriented guilt and identity-oriented shame illustrate how such experiences can overlap and influence each other. This book aims to recognize these phenomena, identify their practical implications both for those who experience them and for their broader world, and articulate an ethical response to them. Along the way, I illustrate how understanding environmental guilt and shame can enrich environmental ethics.

Following moral and social psychologists, I take the feelings of guilt and shame to be self-conscious moral emotions or emotions based on one's judgments about the morality of one's actions, thoughts, and/or one's self (Barrett 1995, 34; Fischer and Tangney 1995, 3; Kitayama, Markus, and Matsumoto 1995, 441–2; Tangney 1995, 114). Guilt and shame are considered inherently relational emotions because the harm upon which they are grounded is harm to another entity in a relationship with the feeling agent. Indeed, relationships and social influences on identity (e.g., through communication, cooperation to meet needs, shared attitudes, role relationships, and the sheer self-identification with a social group) are presumed in this work (Turner 1982, 15–18). Using social psychology, historical studies, and philosophical arguments, I extend these notions of individual guilt and shame to memberships and collectives where the whole is or is not limited to the sum of its parts.

Guilt and shame as self-conscious moral emotions have fairly predictable outcomes. Guilt generally inspires agents feeling it to do better—as when a grocery shopper brainstorms about how to remember her cloth bags before her next trip to the store. Shame, focused as it is on identity, often produces the urge to shrink or hide, as well as despair and paralysis as individual or collective agents experience themselves as fundamentally flawed and therefore unlikely to be able to change. Shame often prompts existential reflection about what it means to be oneself, a part of one's society, or to be human in general. Frequently, the result is a desire for future ethical behavior coupled with knowledge that failing to live up to ideals is inevitable.

1.1 Questions of Environmental Guilt and Shame

Narratives like Beavan's resonate with my own experience and that of dozens if not hundreds of people who have confessed their self-perceived environmental wrongs to me after learning I am an environmental ethicist. After experiencing many confessions and reading many more in popular and academic environmental literature, I had many questions: How prevalent are such feelings in the United States? Are they normatively significant or just self-indulgent and whiney? What are their ramifications for environmental action? Should we feel this way? Should we induce these feelings in others? How should we respond when we or others have these experiences? Why have environmental ethicists and scholars of religion not dealt with them more substantially? As I studied the primary sources, I realized that experiences of environmental guilt and shame and the practical and existential questions they raise are widespread among U.S. Americans today and recognized by many more, even those who do not admit to feeling them. These heartfelt confessions are not always or merely narcissistic navel gazing or trivial problems. Rather, they are significant phenomena for multiple practical and conceptual reasons.

In particular, guilt and shame can help agents be moral. Experiencing them may prompt people to recognize that they are not living up to their ideals, that they hold previously unacknowledged values, or that they are deeply committed to values previously thought to be peripheral. Such realizations can catalyze behavioral and identity change to deal with the problems leading to guilt and shame as well as to avoid experiencing them. Such ethical development is particularly valuable in new or changing circumstances such as those of climate change, where some people may develop new moral commitments or extend old values to new situations more quickly (Isaacs 2011). Online confessions of environmental wrongs are one such constellation of actions. They constitute an emerging, unplanned ritual that help people articulate their values, make sense of what their infractions indicate about their identity, collaboratively determine how to do better in the future, and support others in these endeavors (Fredericks 2014b). People may also listen to those they have harmed, apologize, make reparations, or change behaviors and/or social and infrastructural systems to prevent the problem from repeating.

Guilt and shame do not, however, always have positive consequences. While guilt often motivates people to take constructive action, this very impulse can also leave them vulnerable to manipulation. For instance, advertising campaigns often promote individualistic green consumerism over policy, infrastructure, or industrial changes. This focuses popular environmental action on small, relatively ineffective changes such as the type of bags one uses at the grocery store, rather than changes with a more significant impact, such as minimizing air travel, increasing public transportation, or improving building standards. Shame can involve such a totalizing negative experience of the self that agents are unable to conceive of or hope that they can do or be better. Shameful agents can hide from or be paralyzed by their perceived total corruption, deny the problem, or become so angry that they do not act constructively. This inaction can further exacerbate the actions and modes of existence that sparked the shame, leading to a spiral of ever deeper shame and inaction. Environmental shame thus has consequences not only for shameful agents but also for any humans or other biota harmed by the environmental degradation it engenders and sustains.

Despite these significant consequences of environmental guilt and shame, scholars of religion and the environment and secular or religious environmental ethics have paid them little attention. Many probable reasons for such inattention exist. Religious environmental ethics often seems to presume that environmental wrongdoing falls under existing systems of dealing with human limits and thus does not require specific attention. Despite a variety of calls to take seriously the ethical insights of various groups of people, including women, people of color, and the poor (Jenkins 2013; West 2006), many scholars still give the knowledge of those groups significantly less attention than theoretical ideas or official practices. Additionally, the harms of environmental degradation to vulnerable populations are rightly given more attention than the ramifications of such harming actions for the perpetrators. Finally, emotions are often neglected or rejected compared to rational arguments in environmental ethics such that ethical knowledge informed by emotions is also discounted.

Emotions are not, of course, sufficient guides for an ethic. Some are misguided or based on corrosive or evil ideas (e.g., the emotional impetus of neo-Nazis marching in Charlottesville, WV). Even the best insights stemming from emotions require fleshing out through rationality and experience according to the agent's norms. Attention to emotions should not cause neglect of the effectiveness of environmental actions. While attending to emotions in our ethics comes with challenges, the payoff is worth it. Moral emotions are key parts of human life that inform our experience, concepts, actions, rational analysis, and identity (Jaggar 1989). They are sources of ethical knowledge, and pose ethical challenges and opportunities. Paying attention to them is necessary in order to have an ethic that can respond to these realities of human experience.

Yet, when scholars explicitly acknowledge environmental guilt and/or shame it is often through denials that people have or should have such emotional reactions about the environment, particularly climate change. Climate ethicist Dale Jamieson, for instance, quotes one of psychologist Daniel Gilbert's arguments about why humans are not concerned about global warming:

> global warming doesn't [...] violate our moral sensibilities. It doesn't cause our blood to boil [...] because it doesn't force us to entertain thoughts that we find indecent, impious or repulsive. When people feel insulted or disgusted, they generally do something about it, such as whacking each other over the head, or voting. Moral emotions are the brain's call to action [...] Yes, global warming is bad, but it doesn't make us feel nauseated or angry or disgraced, and thus we don't feel compelled to rail against it as we do against other momentous threats to our species, such as flag burning. (Gilbert 2006, M1)

Jamieson takes at face value Gilbert's conclusions that people do not feel, or do not feel strongly, about climate change, and uses them to bolster his argument that climate change is beyond our existing moral frameworks, particularly of commonsense morality (Jamieson 2014, 168–9). Similarly, Tracy Lynn Isaacs, a philosopher who develops a theory of individual and collective responsibility, dismisses guilt feelings as unnecessary and/or irrelevant because she thinks "objective" standards are sufficient to ground responsibility. Presumably she thinks the emotional experiences of guilt yield no additional ethical content to such an account (Isaacs 2011, 90–92).

When emotions are considered in environmental ethics, it is generally with respect to positive emotions like love or biophilia, though these terms are often considered as virtues to be cultivated rather than as emotions that may be sources of ethical insight (Clowney 2013; Kellert and Wilson 1993; Nash 1991).[2] One of the most sustained discussions of negative emotions in environmental ethics is found in the work of William R. Jordan III and his collaborators on shame (Jordan

[2] In the last few years, an increasing number of environmental studies scholars and the general public have recognized environmental grief, despair, and mourning. While this move significantly expands the consideration of negative environmental emotions, such study is distinct from my project, since grief, despair, and mourning do not necessarily implicate agents as guilt and shame do (France-Presse 2019; Cunsolo and Landman 2017; Foster 2015; Head 2016; Lertzman 2015; Nemeth, Hamilton, and Kuriansky 2012; Orange 2017; Pike 2020; Weintrobe 2013).

III 1993; 2003; Jordan III, Barrett, and Curtis 2012; LeVasseur 2011). They focus on the existential status of shame as much or more than the emotional experience. They are right to note that "An environmentalism that fails to provide productive ways of dealing with existential shame may well prove inadequate to the task of providing means for achieving a healthy, sustainable relationship between humans and the rest of nature" (Jordan III et al. 2012, 291) because existential shame is rooted in people's recognition of their finitude, difference from ideals, and from others. Coming to grips with this is necessary, they maintain, to understand the world to the degree to which we can and imagine and enact alternative ways of being and acting including the sustainable.

Their work, however, needs to be expanded to sufficiently meet these goals. Since they focus on ecological restoration as a form of ritual that can restore nature and people's relationship to it, they do not sufficiently investigate the ways guilt and shame can impact human–human relationships. Doing so is necessary to (1) respond to the impacts of environmental guilt and shame and the conditions that give rise to them and (2) examine the ways that these moral emotions can reveal ideas that are transformative for ethics such as the nascent theory of collectives I identify and then develop.

Outside of ethics, there has been some scholarly attention to environmental guilt and/or shame. *Living in Denial: Climate Change, Emotions, and Everyday Life* by Kari Marie Norgaard and Jennifer Jacquet's *Is Shame Necessary?: New Uses for an Old Tool* treat these subjects extensively. However, given their disciplinary location, in sociology and economics respectively, these authors do not sufficiently address the ethical questions guilt and shame raise. Research on social psychology and environmental marketing using environmental guilt and shame corroborates the existence of these phenomena, as will be discussed in Chapters 2, 3, 5, and 7. Yet, such studies are still relatively rare and are limited in terms of their contribution to this project, as they examine the effects of feeling guilt on taking action on climate change (Ferguson and Branscombe 2010) or briefly mention guilt in a larger study of how one's self-evaluation informs one's support of social issues (Sonenshein, DeCelles, and Dutton 2014, 12) rather than trying to determine the extent, character, and ramifications of feeling guilt and shame in the general population.

With so little discussion of environmental guilt and shame in the academic literature, it is not surprising that many questions remain about these phenomena: How are guilt and shame differentiated? Are environmental guilt and shame philosophically justified? Ethically speaking, should agents feel these emotions? What is the role of the community in eliciting these emotions? Should shame or guilt be intentionally elicited to facilitate pro-environmental behavior? How can individuals and collectives deal with guilt and shame whether or not they are intentionally induced?

1.2 The Ethics of Environmental Guilt and Shame: Collectives and Rituals

In *Environmental Guilt and Shame*, I answer these questions with three arguments, focusing on the first two, which require the most theoretical development: First, not only individuals but also collectives, including the diffuse collectives that cause climate change, can have identity, agency, and responsibility, and thus guilt and shame. Second, some agents, including some collectives, are right to feel guilt and shame for environmental degradation including climate change, though this does not mean that intentionally inducing these emotions is warranted. Third, to conceptually, existentially, and practically deal with the effects of guilt and shame on individuals and collectives and the underlying wrongs that lead to them, a series of safeguards are needed which can be developed and maintained through rituals. Such rituals are already emerging in online and activist communities, but more intentional ritualization is needed to fully deal with guilt and shame and the anthropogenic environmental degradation that may spark them.

To make these arguments, I need to go beyond the moral emotions themselves to metaphysics and metaethics. Confessions of environmental guilt and shame in popular literature often presume a nascent collective identity and responsibility that grounds their guilt and shame.

Much work is needed to elaborate on and clarify these concepts, since Western environmental ethics focuses on individual agency and, to the extent that it considers collectives, considers well-defined collectives rather than the diffuse collectives of the sort that, along with individuals and well-defined collectives, contribute to climate change. This work constitutes the first major argument of the book, *that collectives, even diffuse collectives like those responsible for climate change, can have an existence, identity, and ability to act, and thus responsibility that is not reducible to the sum of their constituents. From this notion of collective identity and agency arises the possibility of collectives being and feeling guilty and shameful.*

Particularly, I argue that neither well-defined and often intentionally acting groups like corporations, nongovernmental organizations, or governments (Isaacs 2011; May 1991; 1996; May and Hoffman 1991), nor loose associations of people in one specific place, such as mobs (May 1987), are the only collectives with identity, agency, and responsibility. Diffuse collectives, such as people in developed nations contributing to climate change, have collective identity distinct from the sum of their individual identities. Arising from this identity is collective agency, as collective ideals, experience (including emotions), social structure, and actions can be different from what the members of the collective would have as individuals. Knowledge about climate change and the causal results of their actions and existence are two of several reasons to consider such groups responsible. While I consider more collectives than Isaacs, I follow her two-part theory of responsibility which maintains that both individuals and collectives may have responsibility in particular situations and that neither type of responsibility is reducible to the other. Considering diffuse collectives as I do does not negate the responsibility of well-defined collectives or individuals.

Developing these metaphysical and metaethical claims while taking emotions seriously allows me to cut through conceptual problems many Western philosophers have in understanding responsibility for collective acts including environmental degradation, particularly climate change. Specifically, their focus on intention and on distinct individuals means that they cannot determine responsibility for climate change in the past or sometimes the future. Jamieson, for example, finds that responsibility is not a helpful ethical framework to respond to climate change because the "probabilistic association between what I do and an increase in harm is not sufficient for my having acted wrongly" (Jamieson 2014, 174–5). Instead, he turns to "green" virtues for individuals as models and motivators for ethical action (186–7), bypassing questions of responsibility. His prioritization of individual agency and direct linear causality prevents a full consideration of the collective causes of responsibility, and thus moves away from this notion which would be useful in the policy arena and our everyday ethical assessments as we attempt to understand and respond to climate change. Isaacs also retreats from past responsibility about climate change (Isaacs 2011, 26, 45, 48, 100, 130), and therefore her work falls prey to many of these same critiques even as her forward-looking collective responsibility is a distinct contribution to thinking about future responsibility for climate action.

On the other hand, many ethicists, especially religious environmental ethicists, assume that people as individuals, and possibly as membership groups (groups where the sum is made of the parts) or collectives, are responsible for climate change to some degree. They do not, however, work out the philosophical details of how they are responsible given the unintentional, cumulative, and complex nature of the causes of climate change. While their work is incredibly important, it is not sufficiently conceptual and will be insufficient practically as long as the main unit of analysis and actions remains individuals. Additionally, insofar as they do not directly deal with guilt and shame or how these emotions affect identity and visions of humanity, they miss important sites of reflection about human existence, limits, and possibilities that are necessary to consistently flesh out their ethics.

My theory of collectives exists between these extremes, preserving the important impulse of each. It seeks philosophical explanations while also attending to applied questions and preserving norms like responsibility that are so integral to our lived ethics. Additionally, I do so in ways that make sense of experiences neither of these groups adequately address—moral emotions, particularly guilt and shame.

After demonstrating that collectives can have the identity and agency necessary to have responsibility and can have collective emotions, it is reasonable to ask whether there are times when individuals and/or collectives should feel guilt and/or shame. My answer constitutes the second major argument of this book: *it can be appropriate to feel moral guilt and shame if the agent in question (a) has standards, (b) does not live up to them, (c) is blameworthy for doing so, and (d) is the kind of being who has feelings about itself.* Both individuals and collectives can meet these criteria. I make the case for an expanded notion of blameworthiness in my first argument. In this second argument, I note that it can be appropriate to have negative emotions including feeling guilt and shame when one does not or cannot live up to one's ideals. It is not appropriate to be primarily jubilant or even neutral in these circumstances, or else one's standards have little to no meaning. Rejecting the validity of experiencing guilt or shame in these circumstances denies part of ourselves. We are not just rational beings. Our emotions help us know the world and ourselves and, indeed, help us become who we are. Guilt and shame are some of the many appropriate reactions to anthropogenic environmental degradation including climate change, because individuals and collectives have done and are doing horrible things to each other, other biota, and ecosystems that will have effects long after we stop the actions. Seas threaten homes as they rise; mild winters enable the spread of vector-borne diseases such as malaria and West Nile virus; changing rainfall patterns hinder agricultural production. Those people and ecosystems who are already vulnerable are often disproportionately harmed by climate change. In these dire circumstances, guilt and shame are not only justifiable but also appropriate reactions for those of us who have participated in systems that perpetuate and exacerbate such damage and injustice.

To be clear, I do not endorse any or every feeling of guilt or shame, but only those that are well-attuned and are based in my expanded notion of blameworthiness that may include individual or collectives. I do not support overbearing or out-of-control consciences that only admit our weaknesses or failings. Rather, we should recognize our failings as well as our strengths, successes, and capacities.

If people should feel some guilt and shame in certain conditions, it is reasonable to ask whether such feelings should be induced and how to deal with them regardless of how they are induced. My answer to these intertwined questions constitutes my third major argument of this book: *Dealing with guilt and shame requires establishing and sustaining conditions to prevent the extreme expressions of guilt and shame and their subsequent harms, a process achieved through rituals.* These conditions include a robust notion of personal, collective, and human nature that allows for ideals and failures simultaneously; systems of social support; and practices which reinforce these and other conditions including having empathy to others, critical self- (or collective) examination, and reform; relationships with others including moral support and exemplars; and effective action. Using the ritual theories of Jonathan Z. Smith and Catherine Bell, as well as Jordan, Joanna Macy, and their colleagues, I argue that these conditions come into being through ritual action as individual and collective action and identity are transformed and sustained. I examine rituals that have emerged to deal with environmental guilt and shame, including online environmental confessional practices and veterans apologizing at the Standing Rock Sioux Reservation prayer camp about the Dakota Access Pipeline for the U.S. military's historic and ongoing treatment of Native Americans, particularly the Sioux. These rituals exhibit some of the characteristics I name but require more attention to meet others including building and sustaining relationships and undertaking effective action to fully respond to guilt and shame and the events which precipitate them.

1.3 Methodological and Conceptual Resources

As I make these arguments, I use methodological and conceptual resources from (1) philosophical pragmatism including learning from doing; recognizing that knowledge, including ethical knowledge, is open to revision; recognizing that facts and values are intertwined; and viewing reality through a pragmatic lens; as well as from (2) religious studies including a capacious definition of religion, and attention to the ways religions are developed, sustained, and revised through the actions of participants.

The phrase "learning from doing" entails that knowledge develops through experience and action whether responding to a problem in everyday life, i.e. how to cope with one's participation in climate change, or reflecting on such experiences to identify ethical recommendations. I reject a priori knowledge. My prioritization of learning from doing prompts me to recognize that the experience of everyday environmentalists can be a source of ethical insights, particularly about collective responsibility. I also maintain that developing and inculcating ideals, identity, and action do and should occur through ritual actions involving bodies, emotions, concepts, community, and often collectives. This method is particularly relevant for dealing with guilt and shame. Guilt can catalyze reparative action, but such initiatives are most effective if guided by reflection and consideration of not only the short-term but also long temporal and spatial consequences. To overcome the degradation of identity, denial, hiding, and paralysis that can occur with shame, agents often need to act differently both to be different and to change their conceptions of themselves. My ethical recommendations will indicate broad parameters of how people can foster such action.

My prioritization of learning from doing also means that ethical values cannot be definitively predefined for all time and all circumstances. Particular values may be useful guidelines resonant with relevant experience, but must always be open to correction and supplementation. For ethical values to work, they must be salient to the situation and their users; allow discrimination between possibilities in complex real-life scenarios, not just in abstract ethical cases; and have a fertile scope such that they can shed light on expansive new situations (among other characteristics). This does not mean that before every ethical decision or action one must forge norms anew, or that I can somehow come to this project with a blank slate, without any ethical priorities of my own that will shape my analysis. It does mean that ethical values and methods must continually be open to evaluation and assessment (Fredericks 2014a).

For instance, this project has come to involve more resonances with virtue ethics over time because the everyday environmentalists with whom I engaged were so concerned not only with the consequences of their actions but also with how their actions reflected their character. Indeed, their concern about their existential status dominates their discussions of environmental shame. Their priorities caused me to recognize more deeply that agents, whether individuals or collectives, need to develop their identity and habits to live an ethical life. Identity formation can occur through a combination of emotional and physical experience, engagement with other people, action, and reflection. The interaction and mutual influence of action and identity is reflected in the interaction of action-oriented guilt and identity-oriented shame, where an agent's actions can often spark shameful reflection about their identity. I often use the term "identity," rather than "character" as is typical in virtue ethics, because "identity" captures not only the characterological dimensions of an agent that develop and change over time but also the way these dimensions arise out of and reflect an agent's metaphysical nature and the degree to which the experience of an agent's character being shaped by particular actions can cause reflection on existential matters.

Experience of environmental issues also calls into question whether our preexisting ethical standards are salient let alone appropriate. Many environmental ethicists have argued, for example, that we need to consider intrinsic values of nature, physical needs of humans, and/or the possibility of environmental injustices more than ethics traditionally has. In this era of climate change, these arguments are joined by questions about what kind of agency is relevant to responding to climate change and who in particular is responsible for it. Predetermining our standards will all too often mean that the individualistic, rationalistic view of human agency and responsibility dominant in modern Western thought will constrain the conversation so much that the moral creativity of people experiencing and reflecting on environmental degradation is overlooked. We will miss insights including the possibility of collective responsibility. Similarly, a hyper-rational ethic will cause us to overlook the existence and significance of environmental guilt and shame for existential reflection and concrete ethical action and hinder us from building tools for responding to our environmental degradation, guilt, and shame.

My understanding of the world and our experience of it, like that of many pragmatists, is not based on a sharp line between fact and value. Rather, I recognize that what we value shapes how we understand the world to be, and how we take the world to be shapes how and what we value even as the world pushes back on our theories and values. Therefore, to do ethical analyses we need to understand the world and values as intertwined. I rely on various kinds of disciplinary knowledge, including from social and moral psychology, ethics, religious studies, and history, to illuminate these connections. I move between these various types of literature as well as that of popular environmental writing to understand the phenomena of environmental guilt and shame, to ascertain what they reveal about us, and to articulate how we should respond.

I employ a pragmatic notion of reality in which the way things function in the world determines their reality. Thus, the emotional experience of collectives along with their relational and biological ramifications as well as their influence on climate change indicates their existence. This view of reality allows us to recognize and deal with collectives without negating the reality of individuals. This view of reality also informs my vision of human nature. I assume a minimal human anthropology in which humans have a host of limits (physical, epistemological, volitional, experiential, etc.) and yet have the capacity to think, feel, and act in potentially positive, constructive ways. My anthropological assumptions are minimal in that they are resonant with and specifiable by people of many different religious or philosophical traditions rather than presuming just one (Fredericks 2014a). For instance, many Christians and Buddhists will agree that agents face limits and have positive capacities.

Analytical categories and research methods from religious studies also inform my work, particularly in my study of everyday environmentalists and my turn to ritual as a means of responding to environmental guilt and shame. I use a broad definition of religion which enables me to consider phenomena classically conceived of as religious, i.e., Judaism, Christianity, Islam, Hinduism, Buddhism, etc., as well as new religious movements, civil religions, and nature religions. They all generally share characteristics such as doctrines, rituals, narratives, emotions, ethics, social structures, and material objects (Smart 1996), though they may not believe in a personal deity or have a sacred text as Westerners stereotypically presume that religions must. These religions, broadly conceived, are cultural systems comprised of symbols that establish "moods and motivations" through an "order of existence" such that the moods and motivations are taken as real (Geertz 1973). Religion understood in this way allows us to look for and learn from a wider variety of human experiences but not necessarily connected to a long-standing religious tradition, e.g., everyday environmentalists.

Such conceptual moves are common in studies of religion and nature, whether about nature religion or religions in general (Albanese 1990; 2002; Taylor 2007a; 2007b), particular environmental groups within traditional religions (Baugh 2016; Finnegan 2011), or the religious dimensions of environmental groups (Pike 2016b; Sanford 2007; Taylor 2010). Elsewhere, I expanded theories of religion and nature to online, ephemeral communities to argue that confessions of everyday environmentalists are religious rituals. These confessions share many characteristics of what Bron Taylor has identified as nature religions, including presuming that nature is worthy of care, that it is problematic to degrade it, and that taking action to live out such care and avoid such harm is an important part of life (Fredericks 2014b, 72; Taylor 2007b, 867). Similarly, their individualism and life priorities align with key features of Catherine Albanese's notion of nature religion (Albanese 1990; 2002). In comparison with other nature religions that have been studied, the everyday environmentalists I examine often have even less structured communities, and, particularly when they deal with their environmental guilt and shame online, may only obliquely reference the environment, as they presume that anthropogenic environmental degradation is problematic but focus on how participating in such degradation makes them feel (Fredericks 2014b, 71–4). Despite these differences, I rely on my earlier work to maintain that the practices of everyday environmentalists align with the category of nature religion (Fredericks 2014b). Analyzing these phenomena with the lens of religion allows us to see the ways in which people develop, are influenced by, and live out metaphysical ideas, ethics, and rituals to constitute their individual and collective identity (Jenkins 2013; O'Brien 2017). These moves are particularly important in this study of the existence of and ways to deal with environmental guilt and shame. Many people experiencing guilt and shame phenomena express and address them outside of traditional religious spaces. Additionally, traditional religions do not (at least yet) seem to adequately acknowledge or deal with environmental guilt or shame. Thus, to understand these phenomena and how people do and can deal with them, we cannot restrict our analyses to traditional religions, even as we recognize that they too may have rich resources for dealing with guilt and shame.

1.4 Values

While my study of environmental guilt and shame arises out of the experience of everyday environmentalists, when it comes time to identify the ethical issues of and responses to guilt and shame, this literature can only take us so far. It is not explicit or detailed enough about its ethics. Additionally, everyday environmentalists' attitudes and practices sometimes exacerbate rather than relieve environmental degradation, guilt, and shame. Consequently, I look to academic literature on guilt and shame in general and environmental guilt and shame in particular to flesh out my analyses, even as academic literature may not capture all of the values of the everyday environmentalists, their prioritization of collectives, or the wide variety of cases of environmental guilt and shame. Recognizing these limits, and that I am also an everyday environmentalist in some respects, I rely on my own values in the ethical analyses of guilt and shame and my constructive arguments about how to respond to them. Of course, my values also influence the whole project to some degree, as no one can be value-neutral. I rely on a set of values including dignity, participatory justice, multidimensionality, responsibility, effectiveness, and feasibility. Here, I briefly describe each of them in turn even as each informs the others, and consequently their descriptions will necessarily overlap.

Probably the most important value for this project is recognizing and respecting the *dignity and value* of all humans. This impetus manifests itself in at least three aspects of the project. The first is motivational. Beyond curiosity about environmental guilt and shame, I am concerned for the many people who are harmed when contributors to environmental degradation do not deal with their guilt and shame in productive ways and therefore perpetuate environmental harm. I recognize the deep disparities in contributions to and harms experienced by environmental degradation and the ways such disparities unjustly belie human dignity and the value of other entities. Recent scholarship and activism for environmental justice notes that such harms do not just consist of economic damages, disrupted ecosystems, or diseases but also includes the loss of the ability to conduct culturally important rituals; weakened community ties; and a host of emotional responses including fear, anger, distrust, and apathy (Figueroa 2006; Figueroa and Waitt 2010; Preston 2011; Strickland and Miller 2007; Verry, Heffernan, and Fisher 2005; Weisman 2012). Responding to such injustices is not only a matter of distributing benefits and burdens or ensuring that all get to participate in decision-making that affects them, but also recognizing and responding to these emotional, relational, and cultural harms as well as the more often acknowledged physical and economic harms.

Restorative justice practices aim to respond to these factors through a diverse set of aims broadly concerned with "replac[ing] our existing highly professionalized systems of punitive justice and control [...] with community-based reparative justice and moralizing social control" (Johnstone and Van Ness 2007, 1). Often restorative justice includes encounters between perpetrators and subjects of harm (Conrad 2015, 26–7) that recognize and address the emotional reactions, harms, and assertions of the subjects of harm through listening, apologizing, reparation, and policy change. Such initiatives include ways of reconciling, making up for past and present injustices, and envisioning and instantiating new futures through recognition, relationship building, and policy change. These initiatives do not work well, however, if those harmed are the only participants. Perpetrators also need to come to the table. If they are too ashamed, they may not be willing or able to participate effectively. If they feel guilt, they may aim for quick solutions to assuage their bad feelings rather than long-term solutions. In either case, ignoring emotions of perpetrators (as well as victims) will almost certainly hinder justice. Therefore, even if one's sole aim is to address environmental injustices, one also needs to attend to the emotional experience of the perpetrator. If one also recognizes perpetrators to be people who deserve consideration, as I do, then helping them deal with their wrongdoing and changing their ways is also important to avoid or at least deal with the devastation of their guilt and/or shame.

My commitment to the dignity and value of all people also undergirds my method of looking to the experiences of individuals and collectives. Namely, I examine popular environmental texts to identify and understand their author's ethical problems and gain insight into ways of addressing such problems. I value the authors as individuals and as participants in collectives. While all experiences can shape an agent's identity by influencing how they act, think, and value, shame is particularly relevant because it stems from how agents conceive of themselves. Consequently, if one dismisses another's experience of shame as insignificant, one in effect dismisses them—a move that runs counter to the dignity and respect all deserve. If done frequently enough or even once with enough force, such dismissal can influence people's vision of themselves, their physical and mental health, and their actions and opportunities in life. Paying attention to the guilt and/or shame of agents does not, however, require a blind, naïve, or uncritical acceptance of all they do or say. At minimum, it requires a willingness to take such experiences seriously.

A third way this project prioritizes the dignity of others occurs as I recognize the danger of shame (on its own or catalyzed by guilt) to erode the identity and value of the individual or collective agent's identity in its own estimation and that of others. This risk, coupled with my concern for dignity, shapes my recommendations about how to respond to guilt and shame by building and sustaining resilient agents through rituals.

My commitment to dignity relates to the value of *justice*. It includes multiple forms—restorative justice, as mentioned earlier; as well as distributive justice, the equitable distribution of benefits and burdens of human interaction with the environment; and participatory justice, the ability of people to meaningfully participate in decision-making that affects them. Because of these and other values, I do not prescribe particular ritual responses to guilt and shame for any or all circumstances, but rather articulate features of rituals that, if performed, will allow agents to address the challenges of guilt and shame in particular and human limitations in general. Participatory justice indicates that the people affected by the ritual should participate in its maintenance and enacting, if not also its creation. Longstanding rituals, e.g., Catholic confessions, may serve this function, as may newly emerging rituals such as the online confessions examined in Chapter 2.

Responsibility also plays a central role in this project. It can be a state of being or the result of a normative judgment. It can be oriented to past, present, or future actions or states. Multiple individual and collective agents may be responsible for a particular event or set of circumstances. Responsibility is often linked to causality insofar as an agent understood to have brought something about is said to be responsible for it. As I will argue, emphasizing the direct linear causality of individual agents, while a key assumption of much modern Western ethics, is limited with respect to climate change and other actions brought about by diffuse collectives. This limited concept of responsibility is problematic because it can obscure the ways in which not only individuals but also collectives are shaped by their actions and need to take responsibility for what they have done or been and to act ethically in the future.

Multidimensionality involves noticing and taking account of all of the salient elements of a situation without overly emphasizing one part.[3] My commitment to this value comes from my presumption that human life, and existence in general, is multidimensional and complex. To understand and make recommendations about a situation requires considering its various aspects. I prioritize the ethical insights gained through emotional experiences of a variety of different people and reflection on such experiences. While scholars who share this commitment often examine the ethics of radical environmentalists, environmental professionals, or religious leaders (James 2013; Johnston 2013; McFarland Taylor 2009; Pike 2016a; Taylor 2010), I primarily study the ethics of everyday environmentalists. I continue this attention to the experience and expertise of laypeople throughout the book, though I also acknowledge that the knowledge of formally trained experts and intentional leadership (of environmental scientists, environmentally conscious policy-makers, activists, ethicists, and religious leaders) is also needed to draw out the insights of popular texts and critique and supplement their ethical insights when necessary.

My commitment to a more comprehensive approach also leads me to recommend rituals as a mode of responding to the various dimensions of being human involved in guilt and shame—emotions, bodies, concepts and rationality, social interaction, and individual and collective agency. In other words, I look to rituals rather than just developing narratives, ideas, or policies because I recognize that the problem is a complicated one involving many dimensions of what it means to be human, and so the solution needs to respond in kind (Jenkins 2013).

Effectiveness complements multidimensionality as it pushes us to ask if a proposed course of action will achieve the desired results with as little monetary, emotional, temporal, physical, and social resources as possible. As importantly, will such actions enable meaningful and significant change or will they just delude people into thinking they have made such change? All too often in our society, actions are evaluated by short-term economic or physical indicators (GDP, air pollution rates, etc.), which, while important, only capture some of what is valued (Fredericks 2014a). This book's emphasis on emotions aims to counter such trends, but I do not want to swing the pendulum too far in the opposite direction. That path can lead to emphasizing feel-good solutions, often at an individualistic level, such that people ignore the political and systemic actions that can have the

[3] Thanks to Colin Weaver for recommending that multidimensionality describes my project better than the previous incarnation of this value, comprehensiveness, which could imply that my study is a complete account of human experience and environmental degradation when guilt and shame are only pieces of the puzzle.

most substantial environmental impact. Rather, my concept of effectiveness includes attention to effectively meeting physical, relational, social, infrastructural, and emotional aims as well as the scale of action.

Throughout this project, I am also committed to *feasibility*. In *Measuring and Evaluating Sustainability*, I use the term "feasible idealism" as an ethical principle that entails that ethical decisions "are implementable yet do not stray too far from ideals" (Fredericks 2014a, 116). During that project, I found that uniting feasibility and idealism had multiple advantages, including guarding against the extremes of practicality and ideals. This unified stance is still important, but here I lean toward feasibility to counter the idealism of much environmental ethics literature which articulates values, principles, or rules for environmental ethics without sufficiently considering if they are implementable or how people should act when they fail to live up to them. While these ethical projects are predicated on a presupposition that humans have done something wrong with respect to each other and the environment, human finitude is rarely given explicit attention when authors develop their constructive recommendations. All too often, ethicists recognize past limits but presume agents can be successful in the future. The inevitability of failure to live up to ideals, the psychological and emotional burden of doing so, and the consequences thereof are not given much (let alone sufficient) attention.[4]

This gap in environmental ethics is particularly problematic in the United States as we do not have a well-developed, coherent, and widely shared vision that accounts for success and failure in action or for the possibility of a flawed existence coupled with the possibility of improvement. We see this lack in public attitudes concerning the rehabilitation of prisoners and drug users in which an offense is often taken as an indication that a person is marred for life, incapable of change, and/or unworthy of assistance. It is also apparent in self-help culture and expectations of social media wherein marketers and peers encourage people to judge themselves as flawed and/or set unrealistic expectations about what is possible such that the person is never really allowed to have flaws, limits, or failures, let alone to account for their inevitability. These characteristics of our society also exacerbate guilt and shame experiences and inhibit the possibility of addressing them productively. Agents are likely to either have too high ideals and ultimately fail to live up to them and be disappointed, or have no significant expectation that one's self or collective is worthy or able to do well, conditions that can exacerbate guilt and shame. To combat these and other negative consequences of inadequate conceptions of and practices for dealing with human existence, a feasible ethic is

[4] Stephen M. Gardiner's *A Perfect Moral Storm* best attends to these limits and emotions with respect to climate change as he investigates pernicious elements of moral corruption that make humans unlikely to recognize or adequately respond to climate change (Gardiner 2011, 301–47). He does not, however, significantly consider guilt, shame, or collectives.

needed. We need conceptual and practical models which recognize humans as beings with capacities and limits of action. We need, not a naked ethic, divorced from what it means to be human, but one that is clothed with visions of human nature including but not limited to failure, guilt, and shame and lived out in robust practices.

1.5 Scope

A study of environmental guilt and shame could examine any number of populations and environmental problems. I center my analysis on the United States and climate change for a host of reasons.

Practically speaking, the substantial body of English-language texts and other media originating in the United States and other Western English-speaking nations about environmental guilt and shame allows relatively easy access to the experiences of everyday people in these nations. Focusing on English-language literature limits challenges of translating texts and standardizing guilt and shame terminology. It also overcomes some of the difficulties involved in studying cross-cultural differences in experiences and expressions of guilt and shame, though of course, English language culture in the United States is not homogeneous and cultural differences about guilt and shame are not absolute. Limiting evidence of environmental guilt and shame to English-language literature, particularly in the United States, also narrows the scope of the study to a manageable size.

While I focus on the United States, the trends I point out are not necessarily limited to this country. If people in the United States—a very individualistic nation, where denial of anthropogenic environmental degradation is common—are experiencing guilt and shame about the environment, then one would certainly expect people elsewhere, where responsibility for climate change is even more widely recognized, to also experience these feelings, likely at greater rates or with more strength. Preliminary evidence indicates that this is the case. Kari Norgaard's *Living in Denial* demonstrate the presence of guilt and shame in Norway (Norgaard 2011). A National Geographic Study of people in seventeen countries found that people in developing nations were much more likely to say that they agree or strongly agree with the statement "I feel guilty about the impact I have on the environment" than those in developed nations including the United States (National Geographic and GlobeScan 2012a, 77). More research is needed. Until that time, my remarks are primarily aimed at individuals and collectives in the United States, especially those that have been significantly involved in anthropogenic environmental degradation, as well as anyone seeking to understand these problems.

Studying the United States in a text about environmental guilt and shame also makes sense because it has been the largest historical contributor to climate change (Matthews 2016), is currently the second largest national emitter of carbon dioxide per year (Boden and Andres 2014; Statista 2017), and is intimately involved in many other forms of environmental degradation, sources of potentially grave guilt and shame. Yet, climate denial is much more prominent in the United States than in other developed nations, a trend that may come in part from anticipated or suppressed shame.

Additionally, people in the United States, especially those who engage in conversation about environmental guilt and shame, seem to be largely middle- and upper middle-class and often white, a group that is both significantly responsible for environmental degradation and, whether they realize it or not, has tremendous financial, political, educational, and social resources to address it.[5] Yet, we face a number of insidious obstacles to making such changes (Gardiner 2011), some of which, including the paralysis and emphasis on the individual that arise with environmental shame, are relatively unexplored. Examining these phenomena is one of the first steps toward limiting them and their effects on both the most vulnerable and the perpetrators.

This focus likely raises two types of objections. First, what about those, like former President Trump who seem to feel no shame about environmental degradation? I begin to address this important question when I consider whether anyone should attempt to induce guilt or shame. I note that such efforts are unlikely to work without significant rapport and trust between the shamer and recipient of shame. Yet, to fully answer this question requires more work than is possible in this volume, and deserves more attention elsewhere. Focusing on those who do already experience guilt and shame is important in its own right. Indeed, the ethical dangers of individual and collective environmental guilt and shame are serious because guilt and shame is felt in part among those who already agree that there is a moral wrong occurring and thus would otherwise support policy change. These moral emotions may be an insidious, hidden barrier to environmental action.

A second objection will likely occur among those committed to justice who are rightly concerned that studying guilt and shame of environmental perpetrators will encourage navel-gazing narcissism among the guilty and ashamed rather than action by perpetrators in collaboration with those most directly affected by environmental degradation. This is certainly a danger of talking about guilt or shame, environmental or otherwise. However, it is primarily a danger if work like mine is taken as a sufficient approach to restorative justice or addressing anthropogenic environmental degradation in general. I do not make that claim. This study is intended to be one of many parts in the growing environmental justice literature. Ignoring environmental guilt and shame will likely exacerbate

[5] Of course, many other people can and do combat anthropogenic environmental destruction.

injustices. Therefore, studying these experiences is necessary, particularly since they are often overlooked.

Environmental guilt and shame are topics that I am well positioned to address. It is often people who share some of my characteristics as an upper middle-class, well-educated, white, cisgender U.S. American of Christian heritage who have contributed most to anthropogenic environmental degradation including climate change and—as the many examples of Chapter 2 illustrate—feel guilt and shame about such activities and circumstances.[6] Simultaneously, we have significant money and political power which can help alleviate the problem and respond to past and present problems. Participating in these problems diminishes the experience of others even as it alters our identity and experience. To have deep and lasting change, we need to not only change laws, infrastructure, and culture but also address our individual and collective identity, values, and ways of living. To do this involves not only envisioning and instantiating positive change but also dealing with what we have done and been.

Most accounts of environmental guilt and shame do not specify which environmental problem or problems the authors are most concerned with (acid rain, species extinction, pesticide damages to human health, climate change, etc.). Similarly, many of my arguments about environmental guilt and shame could apply to a variety of environmental problems, and indeed, to many non-environmental problems. Yet I focus on climate change. It is the largest and most challenging problem facing human society and contemporary ecosystems and biota today. Even when people today consider other environmental and social problems—deforestation, poverty, or racism—climate change lurks in the background, exacerbating these problems and attempts to address them. Studying environmental guilt and shame also yields insights, including those concerning the nature of collective responsibility, that are particularly important to the philosophical and ethical study of climate change and the development of responses to it.

[6] A conversation with Christopher Carter helped me realize the racial dimensions of this project when, after hearing about my book project, he remarked that it was a project about white guilt. I agree. I, however, decided to leave an analysis of the racial dimensions of environmental guilt and shame to a future project given: (1) the lack of attention to it in the writings of everyday environmentalists (though this may be a signal of their white privilege), (2) the lack of race as an analytical category in social scientific research on environmental guilt and shame and (3) my concern to keep an already large project manageable. I encourage further research on the racial dimensions of environmental guilt and shame. I hypothesize that perpetrators' responses to environmental racism will parallel perpetrators' responses to other forms of systematic racism.

1.6 Chapter Overview

In order to make the three arguments outlined above, I need evidence that environmental guilt and shame exist, especially since academic acknowledgement of them has been spotty at best. Chapter 2 demonstrates the existence of environmental guilt and shame in the contemporary Western, developed world, particularly the United States. To present such evidence, I rely on a variety of sources including informal texts (blogs, discussion boards); environmental self-help books and memoirs; and broader indicators of cultural trends including comics, TV shows, newspaper articles, marketing studies, and counseling programs. I find that environmental guilt and shame is commonplace in parts of contemporary U.S. American society. Agents involved in the experience of guilt and shame frequently blur the lines between individual and collective. Responses to environmental guilt and shame often take a religious cast, as people develop confessional rituals that shape their identity and action. Responses to confessions by fellow ritual participants yield considerable insights as to how to theoretically and practically understand and respond to individual and collective responsibility, guilt, and shame, but require development to be fully fleshed out and implementable.

For instance, the primary sources about environmental guilt and shame examined in Chapter 2 do not define or differentiate guilt or shame consistently or systematically explore their implications. Furthermore, these texts presume both individual and collective identity and responsibility, though these concepts are not explicitly or rigorously developed. I stabilize these debates over terminology, and clarify the locus of agency and responsibility for guilt and shame in Chapters 3 and 4 respectively. Chapter 3 constructs a typology of guilt and shame which fleshes out insights of those experiencing environmental guilt and shame through the use of material from moral and social psychology. It particularly relies on work by June Price Tangney and her colleagues, as well as social, and historical studies of collective experiences of guilt and shame after the Holocaust and other atrocities (Branscombe and Doosje 2004a; Isaacs 2014; Isaacs 2011; Tangney and Dearing 2002; Tangney and Fischer 1995).

Chapter 4 builds an account of collective agency, responsibility, and identity that demonstrates the importance of moral emotions alongside rationality. This argument draws on but extends the work of multiple philosophers and theologians including Karl Jaspers, Hannah Arendt, Larry May, and Tracy Lynn Isaacs. Finally, the chapter also explains why in some cases it is not only possible but also appropriate to experience environmental guilt and shame about climate change as an individual or collective.

Since Chapters 3 and 4 move significantly away from the experiences of everyday environmentalists as I build the typology of guilt and shame and articulate its philosophical foundations, it is reasonable to ask how well the theories align with the evidence. Thus, in Chapter 5 I return to the experience of everyday environmentalists to demonstrate that that all elements of the typology are found in accounts of environmental guilt and shame including collective guilt and shame. As collective guilt and shame are likely to be most controversial to readers and have the least evidence in the previously examined data, I will supplement the evidence of Chapter 2 with additional popular and academic literature about collective guilt and shame.

Chapter 6 is a brief interlude from my three main arguments as I explore why other climate ethicists including Jamieson, Gardiner, Isaacs, and J. Baird Callicott have not examined moral emotions or the negative emotions of guilt and shame, why their philosophical assumptions prevent them from doing so, and the advantages of taking these emotions seriously, as I do.

Chapter 7 is the first of three chapters exploring the ethics of guilt and shame. It considers the ethics of experiencing and inducing environmental guilt and shame. Since texts by everyday environmentalists do not explicate their ethics sufficiently to answer these questions, I look to other literature from traditions in the same cultural milieu to identify ethical considerations critical to answering these questions. First, I consider the strongest arguments against shaming: Martha Nussbaum's argument based on dignity and power, echoed in the work of the few environmental ethicists who have considered environmental shaming. Next, I turn to pro-shaming arguments based on the urgent need to act regarding climate change and its effectiveness, particularly as articulated by behavioral economist Jennifer Jacquet. Then I examine how these concerns as well as other ethical issues (including the disproportionate shaming of certain groups, especially women, and the unique challenges of shaming collectives) are manifest in the most prominent types of environmental shaming: environmental marketing and naming-and-shaming campaigns. I conclude that the risks of shaming are so great that it cannot be used lightly, even as its benefits and its apparent inevitability suggests that we cannot completely rule it out.

If guilt and shame are to be used, we need guidelines to ensure that they have the desired, and significant, effects. We also need to be able to deal with them even if we do not intentionally induce them. Chapter 8 addresses these concerns by articulating a set of "conditions" or characteristics needed in the shamer, shamee, their relationship, the broader community, and the physical system in order to ethically respond to or induce guilt or shame. These conditions include relationality; trust in, credibility of, and authority of those shaming; moral support; values; a vision of identity that takes account of human limits and failure while maintaining the possibility of at least partial success; participatory processes; identifying with the other; critical self-assessment and transformation; and action. As maintaining all of these conditions at once is difficult at best, intentionally inducing shame is even more of an ethical challenge than previously indicated. Because these conditions are also needed to respond to guilt and shame, people should foster them regardless of their intent to induce guilt or shame.

Having articulated the conditions to respond to or induce environmental guilt and shame, it is reasonable to wonder how humans could develop such resources. In Chapter 9, I argue that religious rituals have the ability to create and sustain the conditions. To ground this argument, I rely on the ritual theories of J. Z. Smith and Catherine Bell, particularly regarding rites of affliction, which respond to disorder or wrong, as well as William R. Jordan III, Todd LeVasseur, Gretel van Wieren, and Joanna Macy's theories of environmental ritual as a way of responding to negative experiences, affects, and states of being. I expand this work to deal with collectives and relationships between humans, as is critical in cases involving environmental injustices. Turning to examples, I evaluate spontaneous confessional rituals about environmental guilt and shame in popular online confessions and an apology ritual at the Standing Rock prayer camp. Given the limits of these rituals with respect to the conditions, I maintain that it will be necessary to intentionally ritualize and likely use multiple rituals to respond to all of the dimensions of guilt and shame.

The Epilogue summarizes the contributions of studying environmental guilt and shame for environmental ethics and sets an agenda for future research. Addressing both the theoretical and applied ethical insights arising from environmental guilt and shame leads to richer accounts of responsibility and agency that better fit the phenomena and enables more well-rounded approaches to environmental degradation than those that eschew emotions, the experience of laypeople, and/or collectives. These more comprehensive approaches can help people avoid denial, paralysis, and incomplete and unsuccessful approaches to environmental degradation including climate change.

2
Evidence of Environmental Guilt and Shame

To date there has been little research about the existence of environmental guilt and shame, let alone about their triggers, extent, or consequences. Thus, the first task of this chapter must be to gather evidence about these moral emotions. I turn first to popular environmental literature, grouping my findings by media form (e.g., blogs and discussion boards; environmental memoirs and how to be an environmentalist books; and miscellaneous media including TV shows and comics) because different forms often express different aspects of environmental guilt and shame. Next, I turn to other signals of environmental guilt and shame: therapeutic trends, psychological and marketing studies, and journalistic reports on this research. This evidence suggests that environmental guilt and shame are significant parts of the moral experience of contemporary people, in at least the middle- and upper-class, English-speaking, Western, developed world, particularly in the United States. I will argue that they are prevalent enough and can have substantial enough consequences to warrant our consideration.

2.1 Blogs and Discussion Boards

Popular media, especially blogs and discussion boards, demonstrate interest in the phenomena of environmental guilt and shame and the presence of these emotions in at least some segments of society. For instance, in an earlier study, I examined twenty-seven news articles, blogs, and discussion board threads with total of at least 1,012 comments from 432 monikers. These texts used terms including "green," "eco-" or "environmental guilt"; "eco-confession," or "eco-sin." While authors of these English-language texts often used terms associated with religions, particularly Christianity (e.g., sin, confession), as they discuss their self-determined environmental shortcomings, if they mentioned traditional religions at all it was nearly always to disavow them (Asma 2010; Howe 2005; Streeter 2008). The texts focused upon negative emotions arising from daily activities deemed not environmentally sound by the writer—forgetting reusable bags at the store, driving, eating nonorganic food, or using disposable diapers. Authors examined common household activities in a very individualistic way, with little to no acknowledgement that economic, social, or political forces or physical

Environmental Guilt and Shame: Signals of Individual and Collective Responsibility and the Need for Ritual Responses.
Sarah E. Fredericks, Oxford University Press (2021). © Sarah E. Fredericks. DOI: 10.1093/oso/9780198842699.003.0002

infrastructure could influence behaviors or be a place for change. Rather, individuals blamed themselves (Fredericks 2014b).

Such confessions are not, however, one-sided shouts into the void of the internet, but are frequently part of an ongoing dialogue with people concerned about the environment. Elsewhere, I've argued that these interactions are a newly emerging implicit religious ritual (Fredericks 2014b). I will discuss ritual theory at greater length in Chapter 9. For now, I focus on the phenomena. The online confessional rituals generally have three main parts: (1) the confession, (2) a response to the confession by the confessor, and (3) a response by at least one other person who contextualizes the guilt or shame in a larger theory of human identity and/or with respect to the confessor's past and potential actions.

Confessions are often highly emotive and focused on specific incidents in the lives of the confessor:

> It happened again last night. In the midst of washing the dishes, a wave of green guilt washed over me. I had filled the sink with piping hot water and soap that was not phosphate-free, and rinsed some non-organic greens down the sink. The back door was open and the air conditioning was on.
>
> I started to think about all of the eco sins I'd committed in the past hour, the past week, the past month—about all the times I had used paper napkins at dinner, tossed a tin can in the trash and driven 10 blocks to a restaurant when I could have walked. Rather than focusing on all of the actions I do take in an attempt to be a good steward of the environment, I obsess over what I could be doing. It's called green guilt. (Helmer 2005)

Similar sentiments are expressed by Joy on her blog.

> Every time I go to the grocery store, I do all my shopping and get to the register to discover that I've forgotten my reusable bags at home. In one of the greenest cities in America, I consistently commit the environmental crime of needing plastic bags to carry my milk and eggs home. Guilty, guilty, guilty. (Joy 2012)

These quotes illustrate the sheer repetition of activities deemed "sins" in online environmental confessions, whether Joy notes that she forgets her bags "every time" she goes to the grocery store or Helmer recognizes the frequency of her sins. Opportunities for sin involve everyday experiences, generally of household activities—buying groceries, bringing them home, driving, using disposable diapers, heating or cooling one's home, etc. Some confessions highlight sins of work. Kathleen Crislip, a travel blogger, writes, "I've felt guilty for ages about flying around at some cost to the planet…I also sometimes feel sorta guilty for promoting travel to relatively less-invaded spots on the globe" (Crislip 2012). A few discuss holidays, whether vacations to far-flung locations or a typical holiday at home. Blogger Gwen, for instance, reflects.

> The holidays. They're over now (phew), but I'm still feeling the aftershocks. It's time for a confession.
>
> Christmas is about gifts, and this year I bought a lot of them, made a couple (too few), and received even more. There might be a recession on, and Americans might have experienced a leaner holiday this year, but you wouldn't know it by looking at the mountain of loot that I brought home on December 26th. It's worth noting that a few of the consumer goods I received were fairly green—I got a compost bin, some gardening tools, and a gift certificate to buy vegetable seeds from Seeds of Change. I also wrapped the presents that I gave in the paper that I saved last year. But for the most part, this Christmas was a nauseating display of consumption and waste. (Gwen 2009a)

Yet, this visceral reaction is not a one-time event for Gwen. It provides a basis for her entire blog, "Green with Envy, Riddled with Guilt." Describing the blog, she writes

> This blog has a couple of purposes. They are:
> A) an opportunity for me... to promote 'green' behavior;
> B) to vent about my own green living challenges and confess my many eco-sins (Gwen 2009b).

Indeed, terms including "eco-guilt," "green guilt," "feeling bad," "eco-sins," "confession," "anxiety," and "despair" are frequently used by writers of news articles, blogs, and discussion boards to describe their failures to live up to their environmental ideals and their responses to these failures. Most confessions employ highly emotive terms to highlight particular actions of the individual author rather than societal or communal problems.

In such confessions, confessors presume an environmental ideal (e.g., conserving resources) though they rarely mention the ideal itself or grounds for it (e.g., care for humans, other biota, or ecosystems; respect for a creator; acknowledgement of the interconnection of all in the world and therefore the importance of caring for all), let alone explore them in detail. The brevity of such texts, often only a few lines, likely contributes to this trend. When confessors perceive that they have failed to live up to or broken an ideal, they may not only judge themselves but also feel bad and have an urge to confess. Confessors may also judge their whole self to be problematic, an existential move often linked to questioning one's ability to change and/or denial of the problem.

After admitting wrongdoing, confessors usually reveal something more about their context and perceived transgression. They may rationalize their behavior by justifying it as necessary for their work or transportation, or to fulfill societal expectations (Tass 2008; Wadler 2010). Confessors may also try to diminish their sin by proclaiming it offset by their environmentally friendly actions such as hanging clothes to dry, breastfeeding, or using green cleaning products (Tass 2008; Willis 2008).

Confessors regularly acknowledge and work to overcome the harm that can come from too much guilt.

> I realize that focusing on all of my little eco sins—like the fact that I'm wearing a t-shirt that was manufactured in China from pesticide-laden cotton while eating a bowl of non-organic cherries that were imported from Chile—is not productive. I must learn to let go of the guilt.
>
> I'm going to continue using the cold water setting on the washing machine, choosing organic produce when it's an option, turning off the computer when I leave the house and steering clear of paper napkins and bottled water. When I feel those inevitable twinges of green guilt, I'll take a deep breath and let them pass. (Helmer 2005)

This stage of response by the confessor may also include setting goals or action plans for the future:

> New Year's resolutions are about hope and change and moving forward, and it's up to me to wipe my own slate clean. Jan. 1 is about closing out the past, taking a deep breath, and trying again. I'll just have to trust the planet to forgive me when I stumble. (Willis 2008)

Yet even as they work to move on, to do better next time, to live up to their moral ideals through their actions, they desire something else—to *feel* differently, particularly about themselves. Indeed, everyday environmentalists emphasize feelings about what they have done and how their actions influence their sense of self over the practicalities of their infraction. After all, they are not just reading a website, magazine article, or book about tips to "save the environment"; they are posting as a part of a dialogue in which they hope to receive feedback from others. They seem to seek feedback about emotions and their existential questions, craving the validation, response, reassurance, or forgiveness of others. This brings us to the third stage of the emerging online ritual: responses from generally anonymous others that address the confessors' sense of self.

Respondents often offer reassurance or comfort through solidarity as they acknowledge that they feel similarly (HeyAllie 2011; Willis 2008; Young 2012).

They recognize that it is not easy to change, but offer hope that it is possible, especially in small steps (Willis 2008). Respondents attempt to bolster the spirits of the guilty by relating these particular problems to the human condition—that no one is perfect. For instance, Markomd wrote in response to a *New York Times* article, "Green is good…The greener the better. However, we are only humans, not saints or tzaddikim. We rightfully have stewardship for the earth, but we do not have 100% control over ourselves, let alone the earth" (Wadler 2010). They emphasize forgiving oneself and others to build up hope again. Kat wrote in response to the same article, "Reducing consumption, reusing and recycling is a healthier and better way to live your life even if you cannot be 'perfect'" (Wadler 2010). Others attest that though wrong was done, the person is not completely bad; they do other actions which are eco-friendly, or offset an especially bad habit with a better one (Lindsey 2012; Tass 2008; Wadler 2010). With such comments, respondents reassure the confessor that they are not alone, are not completely morally bankrupt, and that their shortcomings and accompanying guilt and shame is a part of being human.

This pattern of confession, self-response, and responses from others is repeated on many environmental blogs and discussion boards (Buck 2013; Campbell 2014; Fredericks 2014b; Wolk-Stanley 2012a, 2012b). While most responses directly address an individual who has confessed, some aim at a wider readership. For instance, Wendy Young was inspired to articulate five tips about how to respond to environmental guilt or shame after reading another author's blog post about being guilty about air travel. Instead of writing directly to that author, Young shared her tips for "find[ing] balance" if one is "trapped in the cycle of feeling guilty for not living green enough" with her blog readers (Young 2012).

Regardless of their target audience, responses emphasize emotions, morality, and existential outcomes over practical tips. Participants in these discussions sometimes ask for or receive practical strategies for changing their problematic behaviors (e.g., tips on remembering to take cloth bags to the grocery store) but emphasize the feelings sparked by such infractions using emotive terms such as "confession," "feeling guilt," "feeling bad," etc. Using language of guilt, confession, and eco-sin emphasizes that using plastic bags is not just a technical problem for these folks, it is an emotional, moral, relational, and existential problem, i.e., a religious problem (Fredericks 2014b, 71–3). Emotional, because the discussion is framed using emotive terms and often emphasizes the participants' feelings more than the material outcomes of their actions. Moral, because by doing or failing to do certain activities they fail to live up to their own ethical standards. Relational, because the wrongs that they do may harm others and may be judged wrong by others or by the individual according to values developed in relation to others. Existential, because these failings cause agents to question their identity, purpose, and meaning in life, and in some cases that of humanity as a whole.

The confessional dialogue and actions that follows enables participants to "attempt to rectify a state of affairs that has been disturbed or disordered" (Bell 2009b, 115). Here, the disorder is something that the individuals perceive to have brought upon themselves by failing to live up to their environmental ideals, or something about their identity that they are able to recognize because of such failure. Writing their confessions and responding to others' confessions allows everyday environmentalists to articulate their ideals; recognize how their actions differ from them; and identify, reflect upon, and, in this case, verbally act out how things ought to be according to their worldview with others with shared values. Indeed, participants often articulate an anthropology of human imperfection coupled with a hope for improvement in one's personal life, a way of addressing the existential issues raised by people who continually fail to live up to their own standards (Fredericks 2014b). Such ideas help frame their intentions for further action and, ideally, the actions themselves. Participants in these rituals also suggest implicitly or explicitly that the values they espouse should be upheld by others. These confessions and responses to them may constitute a community of people with shared values and actions.

Discussion boards and blogs with comments may be particularly useful for everyday environmentalists with family, friends, or physical neighbors with different values. Online, people can find others with similar values and explore and maintain them. Additionally, the rapid interaction made possible by internet communication as opposed to traditional print media may enhance community-building and support for people experiencing existential crises about their environmental impacts. At minimum, relationships are desired means of responding to these feelings; in later chapters I will argue that relationships and communities are essential parts of such responses.

2.2 Books

Self-evaluative emotional and existential reflections about environmental activities are not, however, only found in short online texts. Indeed, a number of books published since the mid-oughts include their author's reflections on admissions of environmental guilt or shame: *How Bad Are Bananas? The Carbon Footprint of Everything*; *Confessions of an Eco-Sinner: Tracking Down the Sources of My Stuff*; *The Economical Environmentalist: My Attempt to Live a Low-Carbon Life and What it Cost*; *A Life Stripped Bare: My Year of Trying to Live Ethically*; *I Killed A Penguin: An Ecological Memoir*; and *Spit That Out: The Overly Informed Parent's Guide to Raising Children in the Age of Environmental Guilt*. The books most focused on guilt include *Bag Green Guilt: 5 Easy Steps: Turn Eco-Anxiety Into Constructive Energy*; and *Mom, Will This Chicken Give Me Man Boobs?: My Confused, Guilt-Ridden and Stressful Struggle to Raise a Green Family*

(Berners-Lee 2011; Harding 2009; Hickman 2005; Pearce 2008; Pleasants 2009; Vaze 2009; von Jessen 2009; Wolf 2010). All written by authors living in the United States, Canada, or the United Kingdom, these texts include personal reflections of the author's attempts to determine and live up to their environmental priorities (e.g., reducing their use of toxic chemicals, water, or energy; or improving the working conditions of people who make and distribute the products they use). The longer format of books and frequently extended research involved enables these authors' arguments to differ from the short online texts discussed above in several ways, even though they still confess their wrongs using emotional language and are concerned about what such wrongs reveal about them.

For instance, authors of these books often explicitly discuss what catalyzed their recognition of their culpability, ontological status, and/or negative feelings. The trigger may be talking with a neighbor, friend, or family member (Harding 2009, 1; Wolf 2010, 7); experiencing a different environmental infrastructure in a new town (e.g., recycling facilities, a walkable neighborhood) (Harding 2009, 7–9, 23–8, 35); reading scientific materials (Vaze 2009); hearing a story of environmental degradation and its effects (Vaze 2009; von Jessen 2009); or increasing exposure to environmental reporting in popular news media (Harding 2009, 1). Regardless of the catalyst, the authors connect scientific research about environmental or social degradation to their actions and emotional reflections upon them. Fritzie von Jessen, for example, describes her reaction to reading an article about how anthropogenic climate change, particularly warming Antarctic temperatures, is threatening penguins. Remembering her recent CO_2-producing activities—driving, "buy[ing] a "cellophane-wrapped steak nesting on a Styrofoam tray," using lighter fluid to start the grill, she asks "Did I just kill a penguin?" (von Jessen 2009, 16–17). After setting a goal to reduce his carbon footprint to combat anthropogenic climate change, Prashant Vaze begins his travel chapter with an exploration of his feelings: "The thought of travel fills us [environmentalists] with longing and guilt at the same time...This chapter is chiefly a conversation I am still having with myself to persuade me to fly less" (Vaze 2009, 287). Vaze calculates his carbon emissions over several years, determining that his two flights to India to see family generated 43 percent of his carbon footprint in 2008 and "counteracted all my other good work" to reduce emissions through changing his food, daily transportation, heat, electricity, and purchasing habits (Vaze 2009, 308–12, 317–18). Emphasizing emotions, Vaze notes how difficult reducing air miles can be, especially if they are "love miles," those racked up when flying to maintain relationships (Vaze 2009, 303).

These authors express many emotions, but the negative outweigh the positive. Rare positive or neutral emotions include feeling "good" (Harding 2009, 91, 120–2, 189; Pearce 2008, 96, 208; Pleasants 2009, 26, 53, 93–7) or joy (Pearce 2008, 6) at environmental successes; hope (von Jessen 2009, 239); or pondering whether one can be happy with a less consumeristic lifestyle (Beavan 2009, 44–5; Vaze 2009, 268). Some authors admit to feeling pride (Hickman 2005, 334–5; Pleasants 2009, 93; Vaze 2009, 83–4) or smugness (Vaze 2009, 81, 241) when they do something good for the environment. More commonly, authors discuss feelings of being "overwhelmed" (Harding 2009, 196, 206; Hickman 2005, 48; Pleasants 2009, 13, 24, 31), responsible for environmental degradation (Pearce 2008, 109–10, 130–1; Pleasants 2009, 10, 16; Vaze 2009, 7; von Jessen 2009, 19, 35, 105, 275) and sometimes, because of their responsibility, feeling like a failure (Hickman 2005, 44, 99), "selfish" (Harding 2009, 21), or disappointed in themselves (Hickman 2005, 116–17). Other common emotions include woe (Vaze 2009, 18), frustration that one has not done enough (Hickman 2005, 125; Vaze 2009, 18, 86), dread at trying to decarbonize the most difficult aspects of one's life (Vaze 2009, 101), "worry" (Berners-Lee 2011, 186; Harding 2009, 2, 7, 12, 69, 127, 133–40; Hickman 2005, 113; Vaze 2009, 164, 273; von Jessen 2009, 21, 154), sadness (von Jessen 2009, 41), feeling awful or bad (Pleasants 2009, 66), and "anxiety" (Hickman 2005, 112–13, 124; Pleasants 2009, 11, 15, 22, 29, 30, 42). Prime among these negative emotions is the constellation of self-conscious moral emotions that hinge on judging oneself negatively for actions that one perceives to be unethical: sheepishness (Vaze 2009, 165), shamefacedness (Vaze 2009, 167), "shame" (Beavan 2009, 37, 204, 222; Harding 2009, 1, 19, 21, 41–5, 47, 49, 50, 53, 58, 67, 71, 73, 75, 132, 144–5, 186; Pearce 2008, 4), and "embarrassment" (Pearce 2008, 4). Within this popular category, "guilt" is the term used most often (Beavan 2009, 36, 66–7, 133, 204, 213; Harding 2009, 5, 41, 45, 47, 67, 69, 125, 146, 175, 190, 195, 206; Hickman 2005, first page of the foreword, 92, 107, 11–13, 117, 123, 146–7, 197, 205–6, 305; Pearce 2008, 30, 132; Pleasants 2009; Vaze 2009, xv, 18, 288, 309, 320; von Jessen 2009, 17, 30, 94, 196–7, 232; Wolf 2010, 6, 29, 42, 50, 130–4), though these authors do not have clear or consistent definitions of guilt and shame within their own work or between their work and that of others. As described in the previous sections, authors experience negative emotions about their actions and their identity or existential status.

The book authors express these negative emotions about a broader range of activities than the online authors. Both groups focus on household actions: eating, heating homes, and everyday transportation. Book authors, however, consider activities including buying new clothing, buying and using electronics, taking out a mortgage, having surgery, and being cremated that have lesser-known environmental impacts (Berners-Lee 2011, 94, 103, 113, 115, 123–9, 160; Pearce 2008, 95–142).

Book authors also recognize the broader social and infrastructural context of such household activities more than their online counterparts. Harding, for instance, acknowledges that infrastructure, formal social mandates, and informal peer pressure all influence her family's habits as they move between cities (Harding 2009). When they moved from Vancouver to Calgary, the lack of recycling options and car culture led her family to increase their carbon footprint. Things changed again when they moved to Australia, where they had facilities to dry their clothing outside, a thorough recycling program, and exposure to local food culture. Moving back to Vancouver, she was struck by their neighborhood's walkability and neighbors' environmental priorities. In all of these places, physical infrastructure, social knowledge, and peer pressure shaped the amount she understood and identified with the environmental movement, participated in environmentally friendly activities, and felt about such activities and herself (Harding 2009, 7–69). Over time she came to a deeper commitment to environmental values. These new priorities were accompanied by negative feelings as Harding felt pressure to live up to her community's high standards, especially as she recognized how often she did not.

Authors of environmental books also more frequently name who is harmed by their actions and consider a wider range of entities harmed. They regularly recognize the ongoing and multifaceted effects of their actions on human health, social or political well-being, and/or particular animals or ecosystems. For instance, Fred Pearce aims to track the human and environmental impact of an average middle-class British person's life. Pearce repeatedly notes that boycotting products made in environmentally degrading ways or using sweatshop labor may harm other ecosystems or the very workers one tries to help as businesses move elsewhere. Thus, he advocates for nuanced positions that take the whole environmental-social system into account (Pearce 2008, 231–64).

Admittedly, many books still take an individualistic perspective about the agency involved in environmental harm, as their main goal is to examine the impact of the author or author's household (Beavan 2009; Berners-Lee 2011; Harding 2009; Hickman 2005; Pearce 2008; Vaze 2009; Wolf 2010). Yet, in contrast with the online conversations discussed above, they frequently and explicitly examine community initiatives as well. The most substantial considerations of community action are typically left until the end of the book (Pleasants 2009, 86–92; Vaze 2009, 329–36; Wolf 2010, 136–7). Authors discuss how and why individuals should get involved in changing laws, social structures, or business practices. Most emphasize that cooperating with others has larger impact than would otherwise be possible, though some also name community support as a necessary element to keep one's spirits up (Pleasants 2009, 86) or get advice about what to do (Hickman 2005).

Indeed, the authors studied aim not only to determine how to live a socially and environmentally sound life but also to live such a life and encourage and help others to do so. They generally aspire to understand the effects of their actions by reading scientific research, consulting experts, or interviewing people who affected by environmental degradation as well as reflecting on their own experiences and resulting emotions. Combining these forms of knowledge humanizes the process of calculating carbon footprints or choosing between household cleaners, making for a livelier read and deeper support for readers with similar struggles.

While only one direction of the author–reader relationship is immediately observable, the authors indicate that fostering a relationship with people who share their dedication to environmentalism is a key to facing their negative emotions and changing their actions. Such connections arose as the public responded to Leo Hickman's blog and *Guardian* articles about his year-long experiment to live ethically (Hickman 2005, foreword, 63, 97) and as authors learned from their families, friends, and environmental experts (Harding 2009; Hickman 2005, 1–17, 117; Pearce 2008, 201–6; Vaze 2009, 136, 142, 154, 293; von Jessen 2009, 279; Wolf 2010). Other authors suggested joining an environmental organization or working with one's neighbors, local businesses, or government to increase solidarity and multiply one's results (Harding 2009, 201–6; Hickman 2005, 307; Pearce 2008; Pleasants 2009, 85–91; Vaze 2009, 18, 248, 326; Wolf 2010, 138). Thus, in these books, the process of dealing with negative emotions about one's actions involves not only examining one's shortcomings and reflecting on one's emotions but also taking concrete, effective actions and, to a certain degree, doing so in a community.

These texts tend to have a vein of optimism, if not significant hope about what individuals or groups can do (Berners-Lee 2011, 2; Pearce 2008, 252; Pleasants 2009, 18). After all, if authors did not believe in the possibility of change then there would be no reason to write a book to catalyze change. Additionally, the logic of shame means that authors are likely to emphasize guilt over shame. If they were strongly exhibiting behaviors typically associated with shame such as denial, withdrawal, hopelessness, or anger, they would probably not be writing books. Yet, embedded within these texts are hints that the authors also have had negative feelings about their identity or character. Sometimes, they use the terms "shamefacedness" or "shame" to refer to a problem with themselves rather than in a particular action (Beavan 2009, 37, 204, 222; Harding 2009, 1, 19, 21, 41–5, 47, 49, 50, 53, 58, 67, 71, 73, 75, 132, 144–5, 186; Pearce 2008, 4; Vaze 2009, 167). At other points they admit to feelings hopeless or paralyzed in the face of the significance of environmental problems and their deep contributions to them (Harding 2009, 196, 206; Hickman 2005, 48; Pleasants 2009, 13, 24, 31). Most, however, do not dwell on these hopeless or paralytic feelings because their aim is to figure out how to do something better and inspire or instruct others to do the same.

A prime example is found in Colin Beavan's *No Impact Man*. As noted in Chapter 1, Beavan wrestles with whether he and all humans are fundamentally problematic or have just done some bad things. He struggles with the meaning of his life (Beavan 2009, 43, 113–15). After reflecting on his grandparents' lessons about avoiding waste and appreciating nature and everyday life he ponders several days of his trash:

> You sit down with your trash and you see your life laid before you on the floor, you see what an archeologist would see when he studies your life a thousand years from now, and you wonder: If life begets life and death begets death, does waste beget waste? If my life begets waste, what does that say about my life? Is a waste of resources a sign of a waste of life?
>
> Half an hour on the floor with my trash and I'd gotten myself stuck in an existential morass. (Beavan 2009, 43)

Yet Beavan tries not to get stuck in existential paralysis:

> It's not that I'm a marred human being who took a wrong turn, or that I've turned bad in the twenty-five years since my grandparents wielded their influence over me. It's not that I'm the lazy ingrate I thought I was. But it may be that as a member of the crew of the huge steamship that is our culture, I had acquiesced to some decisions that caused the whole boat to take a wrong turn, and possibly sink. (Beavan 2009, 40)

Here and elsewhere, Beavan alludes to the temptation to judge himself and humans in general as completely problematic, i.e., as a lazy ingrate, or "failure," not just agents who do some bad actions (Beavan 2009, 20, 64, 190). Yet, he self-consciously returns to his claim that humans and he himself are good, or at least neutral. "I'm an optimist about human nature" (25). Though he sees them as problematic, Beavan asserts "Trash and pollution and greenhouse gases, therefore, are not a result of some total flaw in human nature." Rather they are ramifications of industrialization which does not use a closed-loop manufacturing system and instead acts as if infinite resources, including those for waste processing, were available (65–6). The system, not the species, is flawed.

Authors particularly emphasize deeper emotions and concerns about identity as they reflect upon their intended audience. In *Bag Green Guilt*, Jen Pleasants asks her reader, "Do you feel helpless when you hear that the glaciers are melting, the sea levels are rising, and that polar bears are disappearing?" (Pleasants 2009, 12). Vaze wrote his book "to help people settle into acceptance and find the appropriate response" to climate change instead of "get[ting] 'freaked out' by the enormity of the problem" (Vaze 2009, 9–10). Hickman wrote *A Life Stripped Bare: My Year Trying to Live Ethically* after a year's column in the *Guardian* chronicling his attempts to uncover the impacts of his lifestyle on the environment and other people around the globe. More than six hundred people contacted him who had also "often felt a similar feeling of inertia and guilt about not being able to lead a more ethical life. Some felt they had made a successful switch in their lives, whereas others felt that their efforts had been a miserable, even worthless failure" (Hickman 2005, 7 of forward). Here, Hickman describes his readers' paralysis, a signal of shame, especially when coupled with judgments of the self (being a "worthless failure"). Elsewhere, he quotes from a reader, Vince Hennigan from Atlanta, GA:

> I'm not convinced that all the ethical living pays off, either financially or psychologically. Maybe it's me. Maybe I don't have the "guilt ethic" that's needed. Maybe if I had a more developed social conscience. Maybe I'm hooked for life on good-looking waxed oranges, polished apples and genetically engineered tomatoes. Maybe I don't give a damn (but I really want to). Maybe I'm too damn conflicted thinking about it all. God help me, but I tried. Failure goes down hard. I'm truly just a rabid, wasteful, no-good, torn-with-guilt consumer and I know the Golden Gate will be slammed in my face when all this selfish and careless living finally kills me. (Hickman 2005, 146–7)

Hennigan is feeling bad about a particular action and is moving into the realm of self-judgment, "I'm truly just a rabid, wasteful, no-good..." He even frames this judgment in an ultimate sense with his reference to exclusion from heaven. Thus, we see many of these popular authors writing, in part, to address their readers' self-concept and feelings about it. Authors aim to give them practical advice and convince them that they as individuals or as parts of communities can make a difference, even if they have acted problematically in the past. In these ways, the authors try to move their readers (and maybe themselves) from the paralysis and despair that can accompany the feeling that one is a horrible person to believing one has the possibility to act well, at least sometimes.

These books are not, however, just environmental self-help books, interesting only to people hoping to figure out what to do about their own less-than-ideal environmental actions. They are also indicators of a wider recognition of feelings and judgments about environmental actions and one's identity reflected by such actions. After all, given the financial pressures of the publishing world, so many books including such emotional reflections would not be published if there was no market. Additionally, these books do not merely respond to such interests; they also introduce, reinforce, and legitimate them, creating a milieu in which moral emotions about the environment are typical, even expected experiences.

It is reasonable to ask "expected for whom?" In other words, who is the audience for such texts? What are the demographic characteristics of their authors? As mentioned before, I study English-language materials, focusing on the United States, though texts from other English-speaking developed nations are also considered. The books (and internet texts) I study seem authored by, targeted toward, and responded to by middle- and upper middle-class people, as signaled by their careers, ability to consider often more expensive organic foods, reflection on driving and flying, taking vacations and work trips, and even their access to the internet or pleasure reading and free time to read and post. Book and online authors and online commenters that I studied rarely discuss religion explicitly unless it is to disavow or distance themselves from it (Fredericks 2014b, 70). Race and ethnicity were rarely if ever discussed, probably signaling a largely white audience.

The most striking demographic trend among everyday environmentalists discussing environmental guilt and shame in books or online is the gender disparity, particularly in online communities. Of online texts whose authors' gender seem apparent, the overwhelming majority appear to be by women. Indeed, in hundreds of posts, I can only clearly identify only a handful with clearly male pronouns or names.[1] This is not surprising, since authors and readers who are women dominate lifestyle and homemaking blogs. Online texts discussing environmental guilt and shame, like the environmental self-help books written by women, tend to be written from the perspective of someone in the trenches of managing a household (whether for a family, partner, or herself) who cannot step away from these duties. While a few female authors experiment with trying to green themselves or their household, spending lots of extra time and/or money to do so over the course of a year, such challenges are more often undertaken by men, who appear to be stepping into such roles by choice. They pay more attention to household issues once they are legitimated by their writing careers (Hickman 2005; Pearce 2008; Vaze 2009). In contrast, women tend to tell their stories of attempting to move toward environmental ideals within their everyday life when faced with significant limits to their time, money, and information about environmental issues. Admittedly, a larger proportion of the environmental books I studied were written by men and included more references to male readers than the online sources I examined. Books written by men tend to emphasize quantitative measures of economic and physical efficacy over the emotional experience, authors' reflections on their identity crises, and/or the readers' emotions and identity that are often emphasized by women. I will return to the gender dynamics of this literature in the ethical analysis of inducing shame in Chapter 7. For now, let us return to the compilation of evidence of environmental guilt and shame.

2.3 Other Media

Interest in negative emotions about environmental activities is certainly not limited to the explicitly environmental texts described in previous sections. Newspapers including *The New York Times*, the *Washington Post*, the *Los Angeles*

[1] This method is, of course, an imperfect means of determining gender identity, but is all that is possible given the available information, particularly since this research was conducted at a time in which commenting on different websites required distinct usernames rather than a standardized platform such as Facebook (Fredericks 2014b).

Times, and the *Guardian* have run stories on, framed by, or featuring negative moral emotions about people's environmental actions as well as what such emotions indicate about their identity and possibly human nature in general (Dicum 2007; Fahrenthold 2008; Flesher 2009; Jacquet 2015b; McKibben 2016; Rijnen 2003; Turner 2013; Wadler 2010; Williams 2007). Comics in the *New Yorker* (Hamilton 2008) and, even more regularly, in the *Funny Times*, a liberal monthly paper of comics and satire (Darkow 2016; Mueller 2018; Piraro 2012; Price 2015; Schrivan 2012; Warren 2014), draw on these themes as well. People would not laugh if they did not recognize the behavior caricatured in the comic. Similarly, *Towing Jehovah*, a novel by James Morrow, explores the main character's attempt to assuage his guilt after an oil tanker he captained ran aground, causing a massive oil spill on his watch (Morrow 1995).

While much of this media targets upper middle-class liberals, mainstream media also explore these themes. For instance, in her assessment of images of "Mother Earth," particularly the rise of images of a "hurt Mother," in popular ads, fiction, and movies, Catherine M. Roach, observes "We are now sufficiently aware of the damage we have done to feel guilt, remorse, and fear that we have harmed beyond repair that which we love and which we now know, more than ever before, that we are dependent on" (Roach 2003, 131). Explicit discussions of environmental guilt or shame have been featured on Oprah's website (Kosofsky Glassberg 2008) and the *Today Show*'s website (Mooney and Stikes 2012). On January 15, 2014, *Modern Family*, a highly rated ABC primetime comedy and winner of several Emmy awards for Outstanding Primetime Comedy Series, devoted a third of its episode "Under Pressure" to the feelings and actions of Mitchell, a main character who is an environmentalist who meets a new neighbor who is a more hard-core environmentalist (Bagdonas 2014). The abundant references to negative moral emotions and judgments of actions and selves attest to the experience of and recognition of these emotions in the contemporary United States.

2.4 Therapeutic Trends

Given the widespread presence of such emotional experiences, it is not surprising that a growing number of laypeople, psychologists, and academics are concerned with the psychological effects of anthropogenic environmental degradation. While much of this literature centers on grief and despair (France-Presse 2019; Cunsolo and Landman 2017; Lertzman 2015; Nemeth, Hamilton, and Kuriansky 2012; Pike 2020; Weintrobe 2013), other negative emotions are also explored. For instance, Thomas Joseph Doherty, a practicing psychologist, remarked, "Recently, I have also been interested in ways to help people manage apprehensions about climate change or their ecological footprint, or so-called 'eco-anxiety' " (Doherty 2015; Lewis and Clark Public Relations Department 2010). By "eco-anxiety," Doherty means feeling bad about how one impacts the environment and the ramifications of such feelings for one's own well-being, relationships, and environmental actions. Similarly, Margaret Klein, who has a Ph.D. in clinical psychology, has written a book, *Facing the Climate Emergency: How to Transform Yourself With Climate Truth,* and runs a blog, to "[help readers] to recognize, tolerate, talk about, and think through the deeply frightening, painful truths of climate change." She writes, "I hope to help readers convert their (sometimes latent) feelings of anxiety, terror, and despair over climate change into active social and political engagement" (Klein 2015, 2020). Klein reports being helped by her own approach: "I wake up in the morning thinking about climate change, but I am not thinking about destruction; I am thinking about action and intervention and coalition building... It doesn't feel overwhelming and terrifying anymore. It feels energizing and meaningful. It feels like hope" (Klein 2015). The existence of psychological ramifications of environmental degradation is also borne out by the increasing numbers of academics studying its dimensions, whether in the new journal *Ecopsychology*, at conferences such as the Oregon Ecopsychology Symposium in 2015 and 2016, or in graduate programs such as those at Lewis & Clark Graduate School of Education and Counseling. Similarly, in *Climate Crisis, Psychoanalysis and Radical Ethics,* Donna M. Orange articulates multiple reasons why psychoanalysis can and should respond to environmental crises including climate shame (Orange 2017).

While these psychological trends are rather new, Joanna Macy, Molly Brown, and Chris Johnstone have been working on similar issues for at least twenty-five years (Macy and Brown 2014; Macy and Johnstone 2012). Through their workshops and books, they aim to help people deal with their grief, anxiety, despair, and guilt about their participation in and experience of environmental degradation. Macy and Brown claim to have worked with, or inspired through their books, hundreds of thousands of people over the years (Macy and Brown 2014).

Attention to the psychological dimensions of anthropogenic environmental degradation are signals of the concerns of at least some portions of the public. As the psychological resources for recognizing and dealing with these issues are rather new and usually (so far) studied by their proponents, more research is needed to best understand the prevalence, severity, and implications of and ways to respond to such experiences.

2.5 Research on Environmental Guilt and Shame

Despite growing popular interest, social scientific research on environmental guilt or shame is to date relatively rare and limited in scope. Most common are studies of motivation, particularly about consumer behavior, though research sometimes also examines moral emotions about climate change (Ferguson and Branscombe 2010) or how one's self-evaluation (including guilt) informs one's support of social issues (Sonenshein, DeCelles, and Dutton 2014, 12). Despite the limited scope of this research, it can broaden our understanding of environmental guilt and shame beyond the textual studies above.

Most psychological and marketing research about environmental emotions examines the relationship between individual emotions, behavioral intent, and behavior. Some studies focus specifically on positive emotions (Byrka et al. 2010). Others aim to understand whether positive or negative emotions are more motivational, frequently comparing guilt to another emotion, whether pride (Antonetti and Maklan 2014; Bissing-Olson et al. 2016; Onwezen, Antonides, and Bartels 2013), pride and anger (Harth, Leach, and Kessler 2013), embarrassment (Kaiser et al. 2008), shame (Rees, Klug, and Bamberg 2015), or responsibility (Kaiser and Shimoda 1999). Moral emotions (specifically guilt, pride, or shame) have often been found to influence decision-making (Han 2014), behavioral intent (Graton, Ric, and Gonzalez 2016; Onwezen et al. 2013), and behavior (Rees et al. 2015), prompting individuals to be more environmentally friendly (Bissing-Olson et al. 2016). Other research considers emotions to be drivers of environmental behavior alongside altruism, perceived consumer effectiveness (Akehurst, Alfonso, and Goncalves 2012), self-identity, moral obligation, concern for the inconvenience of environmental action (Bray, Johns, and Kilburn 2010) and/or attitudes, and social norms (Bamberg and Möser 2007; Han 2014; Hines, Hungerford, and Tomera 1986/87; Joshi and Rahman 2015; Kaiser et al. 2008). In short, moral emotions in general and negative emotions in particular are not the only influences on behavior, although, when studied, they are frequently found to be a significant (Han 2014; Kaiser et al. 2008)—if not the most significant—factor influencing environmental behavioral intentions and behaviors (Elgaaied 2012). Such research emphasizes the term "guilt" over "shame" and does not always define its use of these terms. When it does, it tends to follow Tangney's definitions of action-oriented guilt and identity-oriented shame (Graton et al. 2016; Harth et al. 2013). Scholarship may also extend the definition of guilt to include anticipated guilt, the emotion of guilt one expects to have in the future if one does something wrong now (Antonetti and Maklan 2014).

Academic research on environmental guilt has examined a range of behaviors including recycling, reducing water or energy consumption (Bissing-Olson et al. 2016), reducing waste (Graton et al. 2016), buying organic food (Onwezen 2015), buying fair trade coffee (Antonetti and Maklan 2014), reducing energy use or paying tax on emissions (Ferguson and Branscombe 2010), or buying environmentally friendly products in general (Onwezen et al. 2013). Most frequently, study participants are college students or others at a university (Bissing-Olson et al. 2016; Ferguson, Branscombe, and Reynolds 2011; Graton et al. 2016; Harth et al. 2013; Rees et al. 2015; Yang, Jiménez, and Yowei 2015), though individual studies also examined coffee drinkers (Antonetti and Maklan 2014), convention attendees (Han 2014), consumers (Dahl, Honea, and Manchanda 2003), "green" consumers (Barbarossa and Pelsmacker 2016), and teachers (Pruneau et al. 2006). Rarely, other demographic factors such as ethnicity (Yang, Jiménez, and Yowei 2015) or culture (Kaiser et al. 2008) are considered.

Research on environmental guilt and shame has a more international scope than the evidence in previous sections, as scholars have conducted research not only on U.S. Americans (Ferguson and Branscombe 2010; Ferguson et al. 2011; Yang, Jiménez, and Yowei 2015), Australians (Bissing-Olson et al. 2016), and British people (Antonetti and Maklan 2014), but also on Germans (Harth et al. 2013; Rees et al. 2015), Dutch people (Onwezen 2015; Onwezen et al. 2013), Italians (Barbarossa and Pelsmacker 2016), and French people (Elgaaied 2012; Graton et al. 2016). These studies indicate that environmental guilt can motivate behaviors to mitigate climate change (Ferguson and Branscombe 2010), and that lower levels of guilt are associated with more positive feelings toward products advertised in ways that suggest the product can reduce environmental guilt (Jimenez and Yang 2008).

While most research on environmental guilt or shame has been narrowly focused, some, generally by market research firms, has attempted a more comprehensive investigation of the existence of environmental guilt and shame about everyday activities. While these businesses often publish a press release about their findings (Cone Communications 2007; Lexmark International 2009; Mooney and Stikes 2012; Shelton Group 2012), full reports of their research methodologies and results are often either restricted to the original client or only available for a substantial fee. Surveys by media outlets have also examined green guilt, but frequently use informal polls and/or do not specify the exact questions asked in the survey, the demographics of the population studied, or the margin of error (if calculated) when they report their results (Freeman 2008; Mooney and Stikes 2012). Given these limitations, as well as the potential for bias in studies run by self-interested groups, I hesitate to give much weight to any one of these studies. Yet, the existence of so many studies attests to the interest in environmental guilt among corporations seeking to market their products, public consumers of news, and groups with an interest in environmental education. Additionally, the broad trends of such research, finding that between 24 and 79 percent of the respondents report feeling some degree of guilt about their environmental activities, corroborate my claim that environmental guilt is a recognizable feature of the contemporary English-speaking developed world (Bennett and Williams 2011; Cone Communications 2007; Eisner 2013; Freeman 2008; Koch 2009; Kornblum 2008; Lexmark International 2009; Mooney and Stikes 2012; National Geographic and GlobeScan 2012a; Shelton Group 2012).

Other studies about individuals' environmental emotions, however, are more reliable because they are both more transparent regarding their methodologies and demonstrate the soundness of their data collection. National Geographic and Globescan conducted a survey on consumer guilt related to environmentalism each year from 2008 to 2012 (National Geographic and GlobeScan 2012a, 7–8). The survey covered 1,000 people in each of 17 countries (Argentina, Australia, Brazil, Canada, China, France, Germany, Great Britain, Hungary, India, Japan, Mexico, Russia, South Korea, Spain, Sweden, and the United States) with access to the internet at home or in public. The 2012 survey found that from 2 to 21 percent of people strongly agreed with the statement, "I feel guilty about the impact I have on the environment," while an additional 12–24 percent agreed with the statement (National Geographic and GlobeScan 2012a, 77).[2] Interestingly, from 2008 to 2012, people in developing countries were more likely to feel guilty than those in developed nations, though those in developed nations are more likely to think that individuals can make a positive difference with respect to environmental issues (National Geographic 2008, 64–5; 2009, 61; 2010, 70; National Geographic and GlobeScan 2012a, 77, 86–8; 2012b, 4).

A study on the environmental impact of printing covered at least 500 people using "national representative samples" in each of twenty-one countries. Seventy-five percent of those surveyed reported feeling guilty about the manner or amount they print. U.S. Americans felt guiltier about printing than their international peers even though they have less knowledge about the actual impacts of their printing habits. Women surveyed were more likely than men to feel guilty about their environmental behaviors: 79 percent of women claimed they felt guilt from "printing unnecessary pages" as compared to 71 percent of men; 71 percent of women surveyed reported feeling guilty about replacing rather than repairing an electronic device whereas only 63 percent of men did (Lexmark International 2009).

Danyi Qi and Brian E. Roe surveyed a representative sample of the United States to investigate knowledge, values, and emotions related to food waste. They found that 77.2 percent of respondents agreed strongly (~55 percent) or somewhat (~22 percent) that "they feel guilty when throwing away food" (Qi and Roe 2016). These three more robust studies also corroborate my claim that environmental guilt or shame are widely recognizable parts of modern Western culture in general and the culture of the United States in particular.

Additionally, several reliable studies examined whether collective environmental emotions could be induced in controlled settings. Nicole Syringa Harth and her colleagues found that when participants read a vignette emphasizing in-group responsibility for environmental degradation, they were more likely to feel guilty and indicate that they intended to take reparative actions (Harth et al. 2013).

[2] The report does not describe how the term "guilt" was translated or discuss the possibilities of cultural differences in guilt perceptions.

Marleen C. Onwezen manipulated positive and negative private and collective emotions to determine what influences people to buy organic food. She found that "positive private [personal] and negative collective emotions are most effective at guiding intentions" (Onwezen 2015). In other words, personal pride and happiness and collective guilt, shame, and other negative emotions were more likely to influence organic purchases than personal guilt or shame or collective pride. Mark A. Ferguson and Nyla R. Branscombe tested whether and under what conditions people feeling collective guilt for U.S. American contributions to global warming. Participants read scenarios describing global warming as occurring through natural or human causes and as having minor or major effects on U.S. Americans in fifty years. They reported feeling collective guilt about global warming most when it was described as having been caused by humans with minor effects for U.S. Americans (Ferguson and Branscombe 2010, 137). In a follow-up study, people feeling more guilt were more likely to say they would participate in climate mitigation activities (140).

In sum, psychological and marketing research involving environmental guilt and shame about many activities both at individual and collective levels supports my claim that these emotional experiences are common features of contemporary life in the United States. Of course, more research is needed to understand the extent of these emotions in the population and who experiences them, to what degree, and about what subjects.

2.6 Journalistic Reports of Environmental Guilt and Shame Research

Further support for my claim that there is widespread awareness and experience of environmental guilt and shame comes from the fact that when journalists report on surveys about environmental attitudes and behaviors they often confess their own environmental wrongdoing (Dube 2012; Haber 2008; Jones 2013a; Koch 2009; Tim 2010). Some journalists introduce an article with a confession: "I was late. It was cold. I was lazy. It was snowing. So I drove the Subaru two blocks to pick up Sammy from his gymnastics lesson. Halfway there, it hit me like a tiny gold-plated ball-peen hammer: I should have walked... I still feel guilty about driving those two blocks" (Hart 2008). Others respond to the content of the survey, here about "green" New Year's resolutions:

> Having spent 10 years in Catholic school, I have a well-developed guilt complex that covers my own efforts to live a simpler, more eco-friendly life. I often feel I'm not doing enough. Here are my own New Year's resolutions:
>
> 1. Get my green house built (it's been a year in planning!)
> 2. Exercise (this is always on my list.)

3. Bring my own cup when buying coffee and own containers when buying take-out (this could be tricky; I sometimes forget to bring my cloth bags to the supermarket.)
4. Drive less (have to talk to my boss about working more from home; could be tricky.) (Koch 2009)

Journalistic confessions are not only found when the survey itself emphasizes environmental guilt (Haber 2008; Hart 2008; Jones 2013a; Koch 2009; Tim 2010), but may also be present when reporting on "bad" environmental habits, parental environmental attitudes, or environmental concern in general (Dube 2012). Environmental guilt and shame may arise without being intentionally induced.

Of course, not all journalists reporting on environmental issues confess their guilt, but those that do often assume that many readers will have felt such an emotion, ask the reader to recall their environmental guilt, or give advice as if the reader had felt such emotions (Dodd 2012; Kosofsky Glassberg 2008; Lack 2013). As journalistic confessions name and draw attention to the emotions of guilt and shame they may legitimize and even promote these feelings, particularly if the reader looks to the media for information, values, or role models. For instance, parenting and lifestyle media cultivate ideas about the ideal life. If this vision involves eco-guilt, then the reader may come to feel it as well, especially if she or he shares the values of the site. Rebecca Dube's reaction to a survey regarding parental recycling illustrates this process: "Most parents say they recycle the same or more than they did before kids, which is giving me 'green guilt' of my own: Let's just say that life with a 2-year-old has me frazzled enough that I don't always take the time to rinse out containers for recycling and pitch them in the trash instead. Forgive me, Mother Earth!" (Dube 2012).

Reporters confessing their environmental wrongs while seeking some sort of reconciliation returns us to the themes with which we started this chapter. Along the way, we have seen two broad types of experience which often coexist. First, judgments and emotions about actions understood to be an example of a failure to live up to one's ideals (e.g., forgetting one's cloth bags at the grocery store) are often accompanied by an attempt to change one's actions (e.g., determine out a strategy to remember the bags). The second type, especially in the popular literature, involves reflection on how actions reflect one's own or group identity, accompanied by a discussion of how to deal with these existential concerns. Repeated attention to such experiences by blogs, books, TV shows, comics, counselors, social scientists, and journalists suggests that they are recognized features of at least some of the English-speaking population in the contemporary Western developed world, especially in the United States.

The rest of this book responds to these agents' requests for assistance in dealing with environmental guilt and shame and the behaviors leading to them.

3
Typology of Guilt and Shame

The experiences of everyday environmentalists documented in Chapter 2 raise several immediate questions: What does "guilt" mean? How does it differ from shame? To what kinds of agents (i.e., individuals or collectives) do such terms apply? In this chapter, I seek to answer these questions by building a typology of the definitions of guilt and shame, paying particular attention to the agents involved.

Clarifying the meaning of guilt and shame, however, is no simple task. In English, laypeople often have a hard time defining or differentiating the two terms (Fischer and Tangney 1995, 15; Harder 1995, 373; Tangney, Burggraf, and Wagner 1995, 343; Tangney and Dearing 2002, 10–11). The disciplines of religious studies, law, psychology, classics, philosophy, and anthropology frequently use very different conceptions of guilt (Katchadourian 2010) and shame (Nussbaum 2004). Even within one field, scholars often disagree about whether the terms describe subjective emotions or objective states, if and how the terms relate, and how they function for individuals and societies.

In the face of this diversity, I maintain that many characterizations of guilt and shame reflect distinct but not contradictory aspects of these phenomena. Thus, encompassing them in one typology is possible. For instance, guilt and shame can be states of being or emotions. Different kinds of agents (individuals, membership groups, collectives) may simultaneously or separately experience, judge, or influence them. This holistic assessment of guilt and shame is critical to ethically analyzing and responding to them, aims of my later chapters.

As I examine the literature on guilt and shame to build the typology, I rely on many disciplinary perspectives, giving particular weight to conclusions corroborated with multiple lines of evidence. I aim to provide accounts of guilt and shame that resonate with the experiences of the everyday environmentalists described in Chapter 2. To keep my study manageable and applicable to those cases, I focus my attention on guilt and shame in English-speaking developed nations or the larger Western European and North American world. I posit that my typology would be generally applicable to other cultural and linguistic contexts as well, since much scholarship suggests that that there is a shared emotional experience including

Environmental Guilt and Shame: Signals of Individual and Collective Responsibility and the Need for Ritual Responses.
Sarah E. Fredericks, Oxford University Press (2021). © Sarah E. Fredericks. DOI: 10.1093/oso/9780198842699.003.0003

something like guilt and/or shame across cultures.[1] I welcome feedback on and revision of the typology from other cultural, linguistic, and disciplinary contexts.

I begin with what will be a familiar perspective for many in the modern West, the individual. I draw on social and moral psychological research to describe feeling guilty and shameful in section 3.1 and states of being guilty or shameful in 3.2. Next, I use social scientific and historical studies of memberships and collectives to expand the type of agents considered. Finally, in section 3.4 I examine how societal factors and external judgments can shape guilt and shame.

3.1 Individual Guilt and Shame Feelings

Social psychological research on the moral emotions, particularly the work of June Price Tangney and her colleagues (Baumeister, Stillwell, and Heatherton 1995, 255; Tangney and Fischer 1995, xi) focuses upon the internal feelings of an individual experiencing guilt or shame. They understand guilt and shame as self-conscious moral emotions, emotions based on one's judgments about the morality of one's actions, thoughts, and/or self (Barrett 1995, 34; Fischer and Tangney 1995, 3; Kitayama, Markus, and Matsumoto 1995, 441–2; Tangney 1995, 114; Tangney and Dearing 2002, 12–13). Guilt and shame are considered inherently relational emotions because the harm upon which they are grounded is harm to another entity in a relationship with the feeling agent. Relationships and social influences on identity are also presumed in this work (e.g., through communication, cooperation to meet needs, shared attitudes, role relationships, and self-identification with a social groups) (Turner 1982, 15–18).

I follow the social-psychological perspective of Tangney and her colleagues rather than other psychological theories for several reasons. First, it examines guilt and shame individually and in their relationship (Tangney and Dearing 2002, 13). Secondly, a number of methodologically distinct, empirical studies (e.g., interviews and surveys about situations, actions, feelings, and journaling studies) corroborate their general differentiation of shame and guilt (Barrett 1995, 28; Baumeister, Stillwell, and Heatherton 1995, 258; Tangney and Dearing 2002, 23–4, 27, 33–7, 118–19, 121). This research also presumes a deep interconnection between minds, bodies, and relationships with the external world (Fischer and Tangney 1995, 3), which seems to reflect human existence as such. While it prioritizes the experience of guilt or shame in an individual, its recognition of social connections provide a foundation for expanded notions of identity needed to

[1] Cultures may experience and analyze these emotions diversely (Barrett 1995, 38; Edelstein and Shaver 2007, 198–9; Fischer and Tangney 1995, 5, 11–13; Li, Wang, and Fischer 2004) and/or respond differently to the same emotion due to different concepts of identity (Bagozzi, Verbeke, and Gavino Jr. 2003), but this diversity seems to be captured by the broad categories of the typology.

account for guilt and shame about collective experience. Finally, its definitions resonate with my own research about reactions to environmental wrongdoing.

Tangney and colleagues often differentiate guilt and shame by their locus of judgment—either the problematic action or one's whole, global self. They often explain this distinction with contrasting phrases: "Look what I've *done*" emphasizes the particular action; "Look what *I've* done" highlights the self (Tangney and Dearing 2002, ix, 24–5). The action-oriented emotion is deemed guilt and the global self-judgment is shame (10–12).

Other differences between shame and guilt from this perspective include the severity of the emotion, the phenomenological experience of the emotion, and typical reactions to the emotion. Guilt can include repetitive thoughts about one's actions and "remorse, and regret over the 'bad thing done'" (Tangney 1995, 117–18). While guilt can be unpleasant, it seems possible for people feeling it to do something different, as it is about particular behaviors. Indeed, people who feel guilt are likely to want to confront their wrongdoing by confessing to others, apologizing, taking concrete steps to repair relationships or physical conditions harmed by their actions, and/or preventing such actions in the future (Baumeister et al. 1995; Tangney 1995, 116–18). Guilt is generally associated with "motivat[ing] people to behave in a moral, caring, socially responsible manner" (Tangney and Dearing 2002, 1–2). Tangney and Dearing, for example, recount a story from their research about a student who felt guilty after withdrawing from college when her parents were paying. After feeling guilty, she got two jobs and paid her parents back quickly (Tangney and Dearing 2002, 19, 25). Some social scientists see guilt as a mechanism evolved to recognize when one's relationships are not properly cooperative or reciprocal and then catalyze action to rebuild this relationship (Goetz and Keltner 2007, 154–5). Guilt can also be prosocial because one may "regulate their own choices to avoid feeling guilty" (Antonetti and Maklan 2014, 121; Mellers and McGraw 2001) and because guilt can prompt relational repair.

Shame is "generally more painful than guilt," likely because it is about one's whole self rather than an isolated action (Tangney and Dearing 2002, 25). Shame is often associated with the feeling of being small and urges to hide one's head, run away, or deny an issue (Lindsay-Hartz, deRivera, and Mascolo 1995, 278). "Intensely painful feelings of shame," write Tangney and Dearing, "do *not* appear to steer people in a constructive, moral direction. Such intense moral pain about the self cuts to our core" (Tangney and Dearing 2002, 1–2; Wong and Tsai 2007, 211). The logic of shame may undercut hope for change because it is associated with one's whole self. If one is that problematic deep down, then there is no hope of doing better. "Rather than motivating reparative action, shame often motivates denial, defensive anger and aggression" (Tangney and Dearing 2002, 1–2; Wong and Tsai 2007, 211).[2] Insofar as shame provokes these responses, it is much less

[2] Other dangers of shame are discussed in Ch. 7.

likely to promote apologies, cooperation, and compromise that may be necessary to overcome social disruption from one's bad actions. Indeed, social psychologists often consider shame maladaptive for individuals if not society as well, particularly if it "involves supporting unattainable or unrealistic ideals" or "when a person accepts the view of others that a particular way of being is unacceptable when it need not be viewed that way (e.g., viewing a racial characteristic or 'disability' as lowly or terrible)" (Lindsay-Hartz 1995, 297; Tangney and Dearing 2002, 94–5).

On the other hand, some psychologists understand shame as well as guilt as prosocial because they can play a positive role in (a) the development of individuals beginning in childhood and (b) the function of human societies. Behaviors expressing guilt and shame or their precursors emerge in 2- to 3-year old children and increase in complexity over time (Barrett 1995, 46, 50; Ferguson and Stegge 1995, 177; Mascolo and Fischer 1995, 93, 95; Stipek 1995, 238, 246, 248). The capacity to recognize standards of achievement or social norms brings with it the desire or obligation to meet them. Interaction with adults who approve or disapprove of their behavior also seems critical to the development of guilt and shame in children (Stipek 1995, 239–40). Over time, children are increasingly able to evaluate their behavior and self, feeling guilt, shame, or pride. The emotional, behavioral, relational, and sometimes cognitive elements of experience mutually reinforce each other.[3] Indeed, the development of a capacity for guilt and shame co-occurs with and enables the development of a sense of ethics, a conscience, and identity in children (Barrett 1995, 41; Hansen Lagattuta and Thompson 2007) as shaped by biological and sociocultural forces (Goetz and Keltner 2007, 154; Kitayama et al. 1995, 440). In society as a whole, shame can "[signal] submissiveness" and thus help maintain social hierarchy (Goetz and Keltner 2007, 154–5). Taking action to avoid shame may also promote prosocial behavior such as recycling (Alpízar and Gsottbauer 2015). In cultures which emphasize relationship over the individual self, feelings of shame may promote reparative behavior.

These differentiations, and the fact that guilt and shame can help people recognize their previously unacknowledged ethical priorities, are illustrated in Table 3.1. The visceral experience of guilt or shame may alert individuals or collectives to the fact that their actions or selves do not align with their values. If people pause to reflect on this disjunction, they may realize that their previously acknowledged values have applications far beyond what they had consciously recognized, or that they actually hold certain unexamined values (Manion 2003). They may come to realize that they value the life of penguins or people threatened by climate change, and do not think they should contribute to such harm by flying to a vacation or attend a work meeting (von Jessen 2009). Similarly, they may

[3] The necessity of cognitive functioning for shame and guilt in children is highly debated (Barrett 1995).

Table 3.1. Characteristics of guilt and shame

Agent	Guilt	Shame
Individual	• Action-oriented—"I did *that*" • Assess a behavior and its consequences • Short-term • Sparks action, reparations (apology, cooperation, compromise) • Reflection on actions may catalyze the re-examination of values	• Self-oriented—"*I* did that" • Assess one's identity • Lingers, feel deeply • Sparks hiding, covering, withdrawing, anger • Reflection on identity may catalyze the reexamination of values

recognize the flaws in their identity when they feel ashamed at being the kind of person whose daily activities threaten the lives, well-being, and communities of others. Through these experiences, values may be identified, extended to new situations, critiqued, or articulated.

Crisp distinctions between guilt and shame are quite useful: they allow us to consistently acknowledge and consider both moral emotions and their interaction. They illuminate how similar moral emotions can provoke very different actions (e.g., "approach/repair" in the case of guilt and "avoid/self-criticize" in the case of shame) (Ferguson et al. 2007, 337), with significant implications for the self-concept and psychological health of individuals and groups as well as for everyday societal function and social movements. Differentiating shame and guilt in this way also aligns with the experiences of Chapter 2: people seem to experience somewhat separable emotions about both action and identity. Yet, it is important to note that guilt and shame may not be so clearly separated in real-life settings (Harder 1995, 373–7). Individuals may actually experience guilt "fused with shame" in which both guilt and shame are present (Tangney and Dearing 2002, 45, 122) or experience guilt which leads to shame.

Despite these advantages of the social psychological concepts of guilt and shame, there are limitations to this research. It has disproportionately relied on U.S. American college students, a limited cultural, generational, and developmental range of the population (Tangney and Dearing 2002, 35), even though increasing numbers of cross-cultural studies in psychological, historical, and sociological research (as described later in this chapter) indicate that diverse populations consider both action and identity, reparations and denial. Methods of studying the characteristics of, prevalence of, and proneness to moral emotions are often hypothetical and retrospective rather than based on lived experience. Thus, one may wonder how accurately the research depicts regular life. Additionally, the scope of activities studied is often limited to low-stakes experiences such as missing a meeting, rather than considering causing a car accident, participating in a war, or contributing to climate change. Tangney and Ronda L. Dearing justify their focus on such everyday hypothetical scenarios by prioritizing the ability of participants to have experienced or be able to imagine themselves in the tested scenarios (Tangney and Dearing 2002, 41–2). They acknowledge that their studies may be less valid for extreme or unusual scenarios. The evidence of Chapter 2 corroborates their claims about the distinctions between guilt and shame, as it focuses on both lived experience and—especially when about climate change—on high-stakes scenarios. I encourage further study into all of these issues.

Thus, while I recognize the limitations of this research, I find that it is also a useful model of guilt and shame that resonates with the environmental guilt and shame I study. I take the conclusions of social psychological studies not as definitive proof that guilt and shame exist exactly as Tangney et al. describe, but rather as indications of key features of these moral emotions, especially in the English-speaking cultural contexts in the United States in which these studies arise.

3.2 States of Being Guilty or Shameful

While psychologists focus on moral emotions, guilt and shame also map onto an agents' state of being. An agent may have done something wrong and thus be guilty. An agent whose character, identity, or metaphysical status is problematic is in a state of shame. I will use the terms "being guilty" or "being shameful" to designate these states, as opposed to "feeling guilty" or "feeling shameful" for the moral emotions as depicted in Table 3.2. Judgments about being guilty or shameful may occur without feeling guilty or ashamed. For instance, one can recognize that one is guilty of breaking the law to steal food to feed one's family but may not feel bad given the circumstances. One may judge oneself shameful, in a state of moral corruption or taint, without feeling bad about oneself if this judgment is based on metaphysical claims about the status of human nature as inherently sinful, corrupt, or flawed—nothing the individual could help. See Chapter 4 for a deeper discussion of the metaphysics of shame. While emotions and states are often separable, they frequently cause or reinforce each other. Recognizing one's

Table 3.2. Types of guilt and shame

Agent judging or feeling	The entity being judged or the object of the feelings	State of being		Experience of moral emotion	
		Guilty	Shameful	Guilt	Shame
Individual	The self	Being guilty as an individual	Being shameful as an individual	Individual guilt feelings	Individual shame feelings

state of being guilty or shameful can lead to feelings of guilt and shame, and feeling guilt and shame can prompt conscious reflection on one's state of being.

States and feelings of being guilty or shameful are also related in that a judgment that one is guilty or shameful is necessary to feel guilty or shameful, though the judgment may not be conscious. Such judgments and the feelings they allow require a commitment to ethical or social values in order that one can assess one's actions or self to fall short of the standards (Barrett 1995, 25; Fischer and Tangney 1995, 8; Jones, Kugler, and Adams 1995, 314; Lickel, Schmader, and Spanovic 2007, 351; Tangney and Dearing 2002, 31; Wallbott and Scherer 1995, 468). These norms may be social, ethical, or some combination thereof. They may be expressed formally (e.g., in the ten commandments or in a personal mission statement) or be implicit commitments. While sometimes the broken norm is obvious, it may not be. For instance, when omnivores order vegetarian food at a restaurant to avoid shame or guilt from eating meat, the norms of group cohesion or respect for others may be more influential than their dinner companion's values about not killing animals for food. Yet, the person feeling guilt or shame is still committed to a norm, whether or not it is the apparent one.

The requirement that a person has to be committed to the norm they broke in order to feel guilt or shame may provoke critical questions. A critic may well ask, "But what of the victim of rape? Do you really mean to say that if one feels ashamed after this trauma that one must support a notion that being raped implies the victim did wrong? Isn't this blaming the victim?"[4] I reply that a rape victim feeling guilt or shame has internalized norms about who is complicit in sexual acts that provide the benchmark for feeling guilty or shameful. However, this does *not* mean that I endorse this self-assessment or blame the victim. Experiences of shame for unjustified harms to oneself or any aspect of oneself that one cannot help are signs of the limitations of the social system in which the ashamed person lives.

Negative feelings of guilt or shame can, upon reflection, lead to a greater self-understanding and deeper knowledge of one's ethical priorities. In the case of the rape victim, the person may find that their guilt or shame is rooted in an ethical framework where people can be treated instrumentally. These ideas may be in tension with other ideas the victim holds about dignity and worth of humans, self-determination, etc. Therefore, reflection on the guilt or shame feeling can provoke reflection about one's values, including a recognition of the degree to which they are in tension and discernment about which values are most deeply

[4] Thanks to students in my spring 2016 'Collective Responsibility, Guilt, and Shame' class, especially Katherine Sturgill, for raising this question, and to participants at the Midwest Faculty Seminar: The Environmental Humanities at the University of Chicago in April 2016, who further pressed me to clarify this point.

held (Fisher 1984, 194; Manion 2003). Such moves may help people reprioritize values and live them out as more integrated selves.

Of course, not everyone is willing or able to critically assess their values while or after feeling shame. This possibility hinges on several factors. First is the degree to which such shame pervades all aspects of the self. Shame does not necessarily involve the whole self (Bartky 1990, 93; Manion 2002, 79). It can be a core but not the defining feature of identity. If shame only affects only some aspects of the self, one more easily gains critical distance from it. Additionally, the more one's community shapes one's identity, the more difficult it will be to distance oneself from that community's ideals, to assess them, and to reverse or jettison them. Finally, those who experience chronic shame due to being consistently judged as different, other, or deficient from the "norm" (e.g., women judged compared to men) may find it difficult to extricate themselves from shame (Bartky 1990, 93; Locke 2007, 159). Although shame does not necessarily lead to self-reflection, experiencing it may help an agent recognize that they are not what they think they should be—realizations that may also clarify their ideals (Manion 2002, 82; Sullivan 2014, 132).[5]

Thus, I am not blaming the victim. The person feeling guilt or shame is the one judging and committed to a norm. Such norms delineate when agents' actions (guilt) and/or identity (shame) are blameworthy, and thus when they are in a state of guilt or shame. Reflection on such norms and what they indicate about one's state of being can be particularly useful in addressing environmental guilt and shame, as will be discussed in Chapter 7.

3.3 Membership and Collective Guilt and Shame

Recent research as well as political and social movements suggest that shame and guilt feelings and states may apply to groups as well as individuals (Branscombe and Doosje 2004a). To feel shame or guilt about a group to which one belongs or its actions, one needs to identify with the group (though one does not always feel shame or guilt if one identifies with a group or individual in a group that has done a wrong) (Lickel, Schmader, and Barquissau 2004, 36–9; Van Vugt 2009). Additional requirements for feeling guilt or shame about one's group or judging one's group to be in a state of guilt or shame generally proceed in parallel with individual cases. One needs to judge that the group as a whole or individuals in the group (whether formal or informal representatives of the group) have done or failed to do something that is an infraction of norms one holds, and, in the case of shame, that this has affected the identity of the group. Then, due to one's association

[5] Feeling bad about perceived wrongdoing should not always lead one to jettison the norm upon which the judgment of wrongdoing is based.

with the group, one feels guilt or shame about the events (Lickel et al. 2004, 38, 41–3; Rensmann 2004, 173) or judges the group guilty or shameful. Action-oriented guilt about one's group is more likely to provoke apologies, reparations, and behavioral changes than shame about one's group, which is likely to result in denial, anger, or victim-blaming (Branscombe, Slugoski, and Kappen 2004, 30; Lickel et al. 2004, 47). It is possible to feel both guilt and shame about a group, though these feelings "are likely to stem from very different interpretations of the event and their differing implications for people's self-concepts" (Lickel et al. 2004, 43; Mallett and Swim 2004, 66).

Group guilt and shame differ from that of individuals in several important respects. Most obviously, people can have group guilt or shame whether or not they directly acted, whether or not they directly benefit from the acts, and whether or not they were even alive for the original act. This is in sharp contrast to the cases in the previous section wherein the person judging and feeling was also the person who acted against the norms or rules. Secondly, the degree to which an individual identifies with the group influences whether that individual feels responsibility, guilt, shame, or nothing about the group's past actions (Roccas, Klar, and Liviatan 2004, 139–41). Sven Zebel et al. studied Dutch reactions to their nation's (a) treatment of Jews during and after World War II and (b) exploitation of people in the Caribbean during the time of slavery. They found that the more people identified with their nation rather than with those harmed by it, the more likely they were to try to justify the nation's past activities and the less likely they were to judge, let alone admit, that the nation had done wrong. The opposite also held. The more individuals identified with victims of their nation, the more likely they were to admit that the actions were flawed and judge their nation as having done wrong (Zebel, Doosje, and Spears 2004). Similarly, Bertjan Doosje and his colleagues found that strong group identification is likely to undermine people's willingness to accept negative information about their group's past, judge it problematic, and "integrate these aspects into their national identity" (Doosje et al. 2004, 98). In other words, the need to identify with one's group may tip one more towards identity-oriented shame, with its accompanying denial and hiding, than perceiving group acts as merely reflective of actions, not identity. Studies of group guilt and shame also suggest that the division between identity and action is harder to make when the action in question is more severe (e.g., participating in genocide) rather than the comparatively minor infractions of cheating in a quiz that many individual studies of guilt and shame emphasize (Branscombe et al. 2004, 28). Thus, climate change, if taken seriously, may be more likely to provoke identity claims (shame) mixed up with assessments of action-oriented guilt than less severe harms such as littering.

Despite these differences, the categories of the previous section generally hold when applied to groups. Terminological changes are needed to differentiate the agents involved, as illustrated in Tables 3.3 and 3.4. I follow Tracy Lynn Isaacs'

Table 3.3. Characteristics of guilt and shame: individuals and membership groups

Agent and who is assessed	Guilt	Shame
Individual about the self	Individual guilt feelings • Action-oriented—"I did *that*" • Assess a behavior and its consequences • Short-term • Sparks action, reparations (apology, cooperation, compromise) • Reflection on actions may catalyze the re-examination of values	Individual shame feelings • Self-oriented—"*I* did that" • Assess one's identity • Lingers, feel deeply • Sparks hiding, covering, withdrawing, anger • Reflection on identity may catalyze the reexamination of values
Individual about another individual in one's group or the group as a whole	Membership guilt feelings • Action-oriented—"My group (or an individual in it) did *that* and therefore *I* am implicated" • Similar characteristics to individual guilt feelings	Membership shame feelings • Group-oriented—"*My group (or an individual in it)* did that and therefore *I* am implicated" • Similar characteristics to individual shame feelings

terminology of "membership guilt" which encompasses when an individual assesses a group to which he or she belongs (Isaacs 2011, 71–81). Although Isaacs only discusses the state of being guilty, not feelings of guilt or shame as a state or feeling with respect to membership groups, I extend the modifier "membership" to these terms. Thus, I extend the phrases used above to "my group (or an individual in it) did *that* and therefore I am implicated" and "*my group (or an individual in it)* did that and therefore *I* am implicated." I avoid the phrases "we did *that*" or "*we* did that" (Lickel et al. 2004, 355), because in some cases the individual feeling the guilt or shame was not directly causally responsible for the problem, though may have implicitly supported it or failed to stop it. I reserve "we" terminology for collectives as discussed below.

Corroborating evidence for these emotions is found in a widespread and growing trend of recognizing guilt or shame states and feelings for the actions of one's group. Frequently, this is evident where a people in a nation come to recognize its guilt for a past injustice and that some form of restitution and/or apology for the victims is necessary. In *The Guilt of Nations*, Elazar Barkan identifies numerous examples of this trend, despite the difficulty of recognizing guilt of one's group because such recognition can disrupt to political and economic life as well as national identity. He claims they point to a new ethical trend: considering collective

Table 3.4. Types of guilt and shame: individual and membership

Agent judging/ feeling	Object of judgment or feeling	State of being		Experience of moral emotion	
		Guilty	Shameful	Guilt	Shame
Individual	Same individual	Individual guilt	Individual shame	Individual guilt feelings	Individual shame feelings
Individual	Another individual in the group or the group itself	Membership guilt	Membership shame	Membership guilt feelings	Membership shame feelings

justice, responsibility, and victimhood not only in theory but also in practice as legal, political, economic, and moral precedents are set (Barkan 2000, 346).

Barkan examines the best-known and widely studied examples of this trend: post-Holocaust German responses to their actions and those of their ancestors (Assmann and Schwarz 2013; Barkan 2000), U.S. American responses to their internment of Japanese Americans during World War II, Japanese interactions with women they made into sex slaves in World War II, Swiss reparations to Jews for deposits of Holocaust victims, Eastern European treatment of minorities in the post-Communist era, restitution of human remains and sacred objects to Native Americans, restitution and apologies to Indigenous Australians by non-Indigenous Australians, and restitution to the Maoris by New Zealanders (Assmann and Schwarz 2013; Barkan 2000; Maddison 2012; McGarty and Bliuc 2004, 118). Many other studies support Barkan's conclusions, including those contained in Nyla R. Branscombe and Bertjan Doosje's edited volume, *Collective Guilt: International Perspectives*, which demonstrate that many collectives around the world feel guilt or shame for their group's actions whether or not they as individuals directly participated in the action. They define collective guilt as arising "from the distress that group members experience when they accept that their ingroup is responsible for immoral actions that harmed another group" (Branscombe and Doosje 2004b, 3), clearly highlighting guilt emotions. Collective guilt may be felt by men in the United States for gender inequity and their privilege over women (Schmitt, Branscombe, and Brehm 2004), by Dutch people about their ancestor's colonialism in Indonesia (Doosje et al. 2004; Zebel et al. 2004), by early 1990s European Union members for past treatment of Baltic nations (Lasas 2008), and by Italians for fascism (La Rovere 2008). In real-life or laboratory settings, scholars also examine the effects of collective guilt or shame on reparations in Chile (Brown et al. 2008), the way that reparations impact collective guilt felt by men for their privilege over women (Schmitt et al. 2008), British feelings about colonialism in Kenya (Kamau, Giner-Sorolla, and Zebel 2013), Polish responses to the Pope's apology to Jews for millennia of mistreatment (Zylicz and Poleszak 2005), and the responses of contemporary Canadians to early twentieth-century Canadian policies toward Chinese immigrants (Wohl et al. 2013).

This evidence may not, however, be adequately captured by the notions of membership guilt and shame articulated above. Frequently, these studies indicate that individuals are not the only agents who judge and experiencing emotions. The actions the group feels bad about are actions that were not and maybe could not be done by individuals alone. Rather, collectives also act as agents in their own right. The emotions are also collective in that they emerge out of the experience of the group and the experiences of the collective action, the identification of group as one who has done wrong and the emergence of new moral norms and judgments. These emotions are beyond what individuals feel as individuals and seem to be caught by others, transferred, as more and more people witness the guilt and shame, recognize their identity, etc.

Some of the clearest such examples come from Australia where many people take collective guilt to encompass both national actions and national identity, possibly being "rotten to the root" (Maddison 2012, 696). Such assessments have sparked significant debate over multiple years about whether and how to respond. Responses have included both formal apologies by the government and "sorry books" in which individuals may apologize for their own benefit from the settler colonial state and/or on behalf of Australia, regardless of how long they have lived there (Assmann and Schwarz 2013, 60; Maddison 2012). If we take these examples as more than just a turn of phrase, but as an indicator of the phenomenon, then our terminology needs to be supplemented by "collective guilt" and "collective shame" to capture the experiences that collectives can have aside from their constituent parts. Distinguishing between terms does not preclude the possibility that collective, membership, and individual guilt or shame can co-occur or that guilt and shame states and feelings can co-occur, lead to each other, or mutually reinforce each other. This may happen within one particular agent or between agents.

Social scientific research on the collective emotions provides support for my differentiation of individual, membership, and collectives, though they typically use the terms "group-based emotions" for memberships and "group emotions" for collectives.[6] "Group-based" emotions are "emotions felt by individuals on behalf of a social collective or other members of a collective" (von Scheve and Ismer 2013, 408). These emotions arise "due to his or her identification with the group" (Niedenthal and Brauer 2012, 269; Sullivan 2015, 384). Group-based or membership emotions may not be shared by others in the group though they are about the group. Think, for example, of a family member who feels embarrassed for the rest of the family because of their unusual dancing at a wedding. The emotional response is about the family and arises out of its activities, but it is felt by a person, not the group (Barsade and Knight 2015, 22–24). Significantly, the individual feels the emotion about something that the group as a whole or members of it did even though the individual did not directly participate.

In contrast, "group emotion"—and many similar terms including "intergroup emotion" (Christov-Moore and Iacoboni 2015, 9), "the result of mood convergence" (Christov-Moore and Iacoboni 2015, 4), "an emotion held or experienced by a plural subject" (Helm 2014, 47), or "collective emotion" (Sullivan 2015, 383)—are shared by and arise out of the group. In the example of the dancers at a wedding, the joy of being together and exuberance of celebrating and moving

[6] The individual as such receives little attention in this research. Individual emotions are understood as emotions primarily about what the individual in question did, thought, or experienced, even as it is recognized that object, substance, and type of emotions will be shaped by one's society.

their bodies may catalyze such a group or, in my terminology, collective emotions. While some theorists argue that the person catching the emotion needs to be in the "physical presence of others" (Mackie and Smith 2015, 263), others argue that emotions can be caught even when the person is alone and does not experience the conditions which caused the emotion. Think, for example, of the many instances of collective emotions for the horrors of World War II, colonialism, etc.

A distinguishing feature of collective emotions is that the experience is not reducible to the sum of emotions of individuals in the collective even if they interact with each other (Christov-Moore and Iacoboni 2015, 2; Helm 2014, 47; Niedenthal and Brauer 2012, 276). Instead, group emotions arise out of experiences and appraisals of the collective as a collective (Christov-Moore and Iacoboni 2015, 11; Helm 2014, 47; Mackie and Smith 2015, 264; Sullivan 2015, 383) and may be considered an emergent property of the group (Christov-Moore and Iacoboni 2015, 3). Scholars studying group emotions do not necessarily presume that the group has a mind or coherent mental states (Helm 2014, 47) though, as we shall see, some argue that this may be possible (Slaby 2014).

To support their theories of group emotions, Leonardo Christov-Moore and Marco Iacoboni draw analogies from studies of other social creatures to hypothesize about humans. They maintain that "social beings in interaction may produce an overall state which no individual member is consciously aware of but which affects all members" (Christov-Moore and Iacoboni 2015, 4). To illustrate their theory, they point to ant colonies or to humans driving around a traffic circle. In both cases there is no overall coordinating boss, yet the individuals play different roles in a collective action that affects all the members. Similarly, they argue that when a group emotion emerges, it may not be completely cohesive among its members and the beliefs and behavior resulting from it may not be completely standardized (4). There is, however, significant similarity and coordination among group members' emotions and actions to yield a group emotion.

Group emotions can occur through emotional contagion, a "family of psychophysiological, behavioral, and social phenomena" that "can be produced by innate stimulus features [...] acquired stimulus features, and/or mental stimulations or emotional imaginary" (Hatfield, Cacioppo, and Rapson 1994, 4). Emotional contagion can occur through several processes, including a rational process in which people consciously identify with each other and/or have a shared commitment and then imagine themselves in the others' shoes and out of solidarity feel as they imagine the other would (Hatfield et al. 1994, 7–9; Helm 2014, 47). Emotional contagion may also occur through conditioned and unconditioned responses. A son constantly subject to his father's anger may get anxious at the sight of anyone who is angry (conditioned). The loud, high-pitched voice of a nervous person can elicit similar emotions in another person (unconditioned) (Hatfield et al. 1994, 9). A third mechanism, "primitive emotional contagion," is most significant for our discussion of group emotions. It is "relatively automatic, unintentional, uncontrollable, and largely inaccessible to conversant awareness," rather than occurring through rational or conscious thought processes (Barsade and Knight 2015, 23–4; Christov-Moore and Iacoboni 2015, 11; Hatfield et al. 1994, 5; Hatfield, Carpenter, and Rapson 2014, 108–9; Niedenthal and Brauer 2012, 268, 272–3; von Scheve and Ismer 2013, 407). Specifically, primitive "[e]motional contagion occurs though automatic mimicry of the facial expressions, voice, and body movements of others [...]." Such mimicry occurs faster than people can consciously decide and intend to mimic another. It "leads the perceiver to actually feel the emotion, effectively catching the emotion of the other person" (Barsade and Knight 2015, 23–4; Hatfield et al. 1994, 5; Hatfield et al. 2014, 109; Slaby 2014, 38).

For example, the averted gaze or shrinking behavior of someone feeling shame can spark analogous behavior and mood in bystanders (Christov-Moore and Iacoboni 2015, 4). A tightly coupled, well-organized group is not necessary. Scholars have amassed considerable evidence for facial, postural, and vocal mimicry and feedback to support this theory both in highly controlled laboratory experiments (e.g., where people's facial expressions were physically controlled) and in more natural settings (e.g., by observing students watching a comedy).[7] Scholars posit that emotional contagion occurs as mirror neurons fire when one experiences an activity *or* sees another experiencing the same thing (Bastiaansen, Thioux, and Keysers 2009, 2391–2, 2396; Blair and Fowler 2008, 310–12; Christov-Moore and Iacoboni 2015, 14; Fourie et al. 2014, 213; Gallese 2001, 35–9; Gallese and Goldman 1998; Gilead et al. 2016; Rizzolatti 2005, 419).[8]

Many if not all of us have first-hand experience of participating in a collective emotion. TV comedies using laugh tracks rely on this phenomenon; producers know a show is often funnier when viewers hear someone else laughing so they give us the opportunity to catch the laughter (Christov-Moore and Iacoboni 2015, 13). Contagious emotions are not, however, always fun. As a graduate student in religious studies, I proctored exams for the law school. Although I did not know any students there and had no particular connection to the students aside from being a student myself, I often could feel and sometimes caught the anxiety present in the room, especially at the start of an exam. Scientists have documented the emotional contagion of a wide variety of emotions including empathy, disgust, guilt, shame, embarrassment, and pride (Bastiaansen et al. 2009, 395; Blair and Fowler 2008; Fourie et al. 2014; Gallese 2001; Gilead et al. 2016; Nakagawa et al. 2015; Preston and de Waal 2002). It is a widespread phenomenon occurring in animals other than humans as well as in humans across many cultures, times, and

[7] See, e.g., reviews in (Hatfield et al. 1994, 53–78) and (Hatfield et al. 2014, 109–13).

[8] The implications of mirror-neuron research for emotional contagion are still controversial (Hickok 2014, 26), but as I recognize that science is always open to revision, I do not take this or any research as definitive. Rather, it is one of many ways the typology is corroborated.

circumstances (Hatfield et al. 2014, 114–19). It has been documented in infants, in kids catching parents' emotions and vice versa, in therapists catching clients' emotions, and in the interlocutors or housemates of a gloomy person catching their mood (Christov-Moore and Iacoboni 2015, 16; Hatfield et al. 1994, 79–81, 83, 88, 102). Emotional contagion also occurs in large collectives, including among 2006 German World Cup fans, participants in the Arab Spring, and among both Norwegians and people with connections to Norway after the mass killing at a summer camp in 2011 (von Scheve and Salmela 2014, xi).

With knowledge of emotional contagion that often occurs automatically and prereflectively, we can return to our question of whether collectives can be the bearer of emotions. To make the shift to collectives, we should notice a potential limitation of the phrases "catching emotion" and "emotional contagion." These phrases can bring to mind an image of one person catching something thrown by another or one person giving a disease to another. In these visualizations, the transfer is unidirectional and between discrete individuals. But emotional contagion seems more complex than this, because passing it along does not mean you are rid of it and emotions do not necessarily come in discrete units. They can be amplified as they are passed along. For instance, consider a recent sitcom plot in which a group of three children stayed home alone for the first time in the evening.[9] One heard a bump and was scared. Others dismissed it. As the sounds continued and the children talked about the sounds and their emotions, they fed each other's fears, becoming so frightened that they ran to hide in a closet. This emotion of fear came to be a collective emotion as the siblings coalesced into a fearful unit. The fear was not just felt by discrete individuals feeling similar while in proximity to one another, but also felt as a collective. There is a way in which there is a collective that was feeling on its own, independently from what individuals would have felt on their own. This occurs when there is feedback among the collective such that the emotional contagion is a dynamic, ampliative, reinforcing process rather than a discrete passing of emotion like a ball thrown from one person to another. This feedback process enables the collective as collective to emerge as emotions grow and sometimes develop beyond what any individual would feel on their own. Here the emotion and the group experiencing it is more than the sum of its parts.

Jan Slaby's expansion of extended-mind theories suggests a way to conceptualize this collective emotion. At the core of extended-mind theories is the idea that "some of our mental processes are physically realized in part by structures or processes in our environment" (Slaby 2014, 32). Slaby suggests that, "for example, tools such as computers and notebooks" are a part of the extended mind when a person interacts with them reciprocally in a way that "reliably leads to enhanced

[9] Unfortunately, I saw this episode several years ago and cannot remember the show.

cognitive performances—ones that the agent on its own would be incapable of carrying out" (Slaby 2014, 32). For example, writing can help one develop new ideas. Slaby develops this theory in two ways. First, he suggests that emotions as well as thought processes may occur through extended mind. Objects, language, and institutions may trigger and expand emotions, as when a thoughtful gift makes the giver's care visible and catalyzes comfort or joy in the recipient. Secondly, Slaby expands the possible elements of the extended mind to include other people and their social interactions. Thus, he suggests that there can be a locus of emotion outside of the individual that gives rise to emotions the individual would not have on his or her own. We recognize that certain people may facilitate particular emotions when we decide whether to seek out our calm friend or our friend who commonly feels righteous, activist anger when we have a problem. While Slaby appropriately acknowledges that extended cognition theory itself is controversial and that his extensions of it are likely to be more so (Slaby 2014, 35), the evidence for such a phenomenon suggest it is as least a plausible possibility. If so, then it may be the case not just that individuals are the repositories of particular emotions, but that collectives are such that the collective as a whole may be said to hold or experience an emotion.

Complicating these analyses of collectives is the fact that the individual and collective identities seem to both be operating at the same time—it is not that the collective displaces the individual or that the individual is fundamental and the collective illusory. Rather, each occurs simultaneously. While we might wish for conceptual crispness, such neat divisions do not seem to map onto the experience or our linguistic capacity to describe them.

Often, we think of collective emotions occurring in co-located groups, as in the example of the frightened children, or in the quintessential examples of group emotions, mobs gathered to protest political regimes or to celebrate or mourn a sporting event. Out of the mob, new emotion and action can occur beyond that of individuals. I maintain that similar kinds of collective emotion can occur with respect to environmental concern. In environmental cases, the emotional tenor is in the air, subtly and sometimes not so subtly, spreading through the population concerned with the environment. Documentaries, formal education, and protests call attention to the horror of anthropocentric environmental degradation. Ads suggesting that parents need to buy nontoxic cleaning products to keep their children safe, nonprofit campaigns shaming corporations (and consumers eating their products) for using unsustainable fishing practices, family members asking innocently (or not) whether the food at the party is organic may all catalyze such emotions, as can one person expressing guilt or shame to another. These catalysts can not only blame individuals but also "all of us," forming a new, albeit vague collective of those in whom the stimulus resonates. Environmental guilt and shame can spread and be amplified between people. They can emerge out of interactions between people even if they are not physically in the same place. Such collective and individual emotions may spark reparative action as well as denial, hiding, and paralysis.

Given the social psychological, historical, and neurobiological research coupled with that of everyday experience, it seems reasonable to say that collective emotions in general and collective guilt and shame in particular exist and can be described through extensions of the terminology above, as illustrated in Tables 3.5 and 3.6. Yet, I realize that the widespread presumption that individuals are the agents and units of analysis in much of modern Western thought will leave many skeptical about the collective part of the typology. To tease out whether and how collectives can have experiences separate from individuals and whether and how they can be judged guilty or shameful apart from their constituent parts, we need to examine whether only individuals or also memberhips and collectives can have identity, agency, responsibility, and emotion. The next chapter will take up this task. First, we need to consider the last dimensions of the typology of guilt and shame: the effect of other people, including one's society, on one's judgments and experiences of guilt and shame.

Table 3.5. Characteristics of guilt and shame: individual, membership, and collective

Agent and who is assessed	Guilt	Shame
Individual about the self	Individual guilt feelings • Action-oriented—"I did *that*" • Assess a behavior and its consequences • Short-term • Sparks action, reparations (apology, cooperation, compromise) • Reflection on actions may catalyze the re-examination of values	Individual shame feelings • Self-oriented—"*I* did that" • Assess one's identity • Lingers, feel deeply • Sparks hiding, covering, withdrawing, anger • Reflection on identity may catalyze the re-examination of values
Individual, about another individual in one's group or the group as a whole	Membership guilt feelings • Action-oriented—"My group (or an individual in it) did *that*" • Similar characteristics to individual guilt feelings.	Membership shame feelings • Group-oriented—"*My group (or an individual in it),* did that *and therefore I am implicated)*" • Similar characteristics to individual shame feelings
Collective about the collective itself	Collective guilt feelings • Action-oriented: "We did *that*" • Similar characteristics to individual guilt feelings.	Collective shame feelings • Collective-oriented: "*We* did that" • Similar characteristics to individual shame feelings

Table 3.6. Types of guilt and shame: individual, membership, and collective

Agent judging or feeling	The entity being judged or object of the feelings	State of being		Experience of moral emotion	
		Guilty	Shameful	Guilt	Shame
Individual	Self	Individual guilt	Individual shame	Individual guilt feelings	Individual shame feelings
Individual	Another individual in the group or the group itself	Membership guilt	Membership shame	Membership guilt feelings	Membership shame feelings
Collective	Collective	Collective guilt	Collective shame	Collective guilt feelings	Collective shame feelings

3.4 The Social Aspects of Guilt and Shame

So far, our examination of guilt and shame has focused upon an agent judging and experiencing its own guilt and shame. Yet, this does not occur in a vacuum. Experiences and judgments of guilt and shame are culturally conditioned and often catalyzed by other agents. This section analyzes those external dimensions of guilt and shame to expand the typology. I begin with the most famous work on the subject, anthropologist Ruth Benedict's *The Chrysanthemum and the Sword: Patterns of Japanese Culture*, though I ultimately move beyond her analysis for a host of reasons.

In *The Chrysanthemum and the Sword*, first published in 1948, Benedict argues that the United States is a "guilt society" emphasizing individual activities, whereas Japan is a "shame society" prioritizing that individual and group action ultimately reflect upon the group or society as a whole (Benedict 1967). For Benedict, guilt depends on "an internalized conviction of sin," whereas "true shame cultures rely on external sanctions for good behavior" and "criticism" or "ridicule" for poor behavior (Benedict 1967, 223). The public awareness of the deed in her notion of shame contrasts with the absolute standard in her notion of guilt. Indeed, she maintains of a person in a shame society, "so long as his bad behavior does not 'get out into the world' he need not be troubled" (223). While Benedict usually considers the actual external community as operative for shame, she does note that the external community can be imagined by the shamed, or approximated by examining oneself (223, 251, 288–9). Given the public/private differentiation here, Benedict argues that while guilt can be assuaged through confessions or "secular therapy," shame cannot (223). Since Benedict prizes individualistic democratic values, she finds guilt more positive than shame. Shame cultures are often perceived, if not by Benedict then by her readers, as less morally developed (Creighton 1990, 279–81).

Benedict's work has been the subject of much scholarly critique in the United States and Japan. Methodological critiques include that she had not studied Japan before beginning her work with the Office of War Information, did not know Japanese, never went there, and did her research in internment camps with Japanese Americans rather than representatives of Japanese society as a whole (Boles 2006, 32; Iwasawa 2012, 360–1; Yoneyama 1999, 75), which are limitations she acknowledges (Benedict 1967, 3–19). Scholars also criticize Benedict's presuppositions, including that Japan is hierarchical, aggressive, or foreign (Boles 2006, 54; Iwasawa 2012, 348, 361; Yoneyama 1999, 71) and that each nation is homogeneous (Bennett and Nagai 1953, 406–8; Boles 2006, 37; Yoneyama 1999, 76). Finally, multiple scholars criticize Benedict for insufficiently corroborating her conclusions (Barrett 1995; Tangney 1995, 136; Tangney, Stuewig, and Mashek 2007, 25).

Despite the criticisms, Benedict's work has been incredibly influential. The book was quite a success in the United States; eight U.S. American editions were published by 1952 and some 350,000 copies sold by 1988. It also garnered significant attention in Japan, where it sold 2,300,000 copies by 1999 (Kent 1999; Pinnington 2001). Benedict's work, including interpretations strongly differentiating between guilt and shame societies, influenced both the popular imagination and scholarship about the relative importance of guilt in modern Western cultures. Her theories have shaped scholarship in anthropology, moral philosophy (especially moral relativism), religious studies, history, philosophy, social psychology, sociology, and Japanese studies.

Popular readers and some scholarly interpreters of *The Chrysanthemum and the Sword* often assumed a stricter dichotomy between guilt and shame than she did (Benedict 1967, 222). Interpreters of her work frequently link the absolute standards of guilt to the monotheism of Western cultures as opposed to the multiple deities of many traditional Japanese religions. They also characterize guilt societies by their individualism—guilt centers on how one's actions will lead to individual punishment. Shame societies are understood to be more hierarchical with formal relational structures, such that one person breaking rules reflects problematically on the whole.

Overemphasizing these differences obscures the fact that while some societies or subgroups may emphasize either guilt or shame, these categories are far from absolute (Barrett 1995; Fessler 2007, 186; Wallbott and Scherer 1995, 467). For instance, many teachers in the United States prepare their students for a field trip by reminding them that they represent their hometown and need to behave so as to avoid denigrating the whole community—a sure signal of concern about how one's actions reflect on community identity.

Tangney et al. take a strong stand against the legacy of Benedict's research, as they find the internal/external differentiation between guilt and shame irreconcilable with their own research (Tangney et al. 2007, 25). However, mandating a choice between these options falsely limits the range of phenomena included in studies of guilt and shame. The study of social aspects of guilt and shame enables the expansion of our typology to include the source or instigator of guilt or shame and the influence of societal norms.

External judgments of guilt include a variety of familiar cases: teachers ascertaining that a student has plagiarized, juries concluding that an agent committed a crime, and the international community determining that a nation has committed genocide. In each one, the judging agent assesses whether another agent lives up to or breaks a normative or legal standard. External agents may punish such behavior or use incentives or education to limit its occurrence.

External judgments of shame can also occur. Consider, for instance, an individual or community judging that a murderer has a flawed character, or environmentalists judging the oil industry to be morally bankrupt. Sanctions for being shameful can include reputational harms, economic sanctions, prison time, ostracism, exile, and excommunication. Alternatively, societies may reward behavior deemed positive through increased status or opportunities. These external judgments and pressure may have a significant impact on an agent whether or not the agent agrees with the judgment that they have acted or been wrong. Such judgments alongside or instead of internal assessments can alter an individual. This change may not penetrate to their most fundamental worth as a person, but it can still affect where, how, and if they live, which in turn can change the person's possibilities, self-concept, and experience. These external influences are not limited to individual agents; they can also occur to and by collectives.

One may wonder whether it is necessary to add these categories of externally judged guilt and shame, as illustrated in Table 3.7, to the typology, since I already have articulated categories of states of being guilty or shameful. After all, internal and external judgements may align. I distinguish these positions, however, because they may differ, and I seek precision in categories. Additionally, without ultimate metaphysical certainty (which I do not think we can have), judgements of states of being will always be contingent. Being clear about who makes such judgments can help people understand why such judgments are made and what their implications may be. Such information is critical in the ethical evaluation of the intentional induction of guilt and shame as discussed in Chapter 7.

Societal visions of whether guilt and/or shame are positive or negative experiences also contribute to how guilt and shame are judged and felt. While the social aspect of the moral emotions is presupposed in the analyses by Tangney et al., I find it important to explicitly name this as a part of the typology to emphasize that agents do not judge or experience guilt and shame apart from their larger context. Thus, we should recognize, for instance, that guilt and shame are seen as completely negative emotions in large swathes of contemporary U.S. American society. Emotions that are uncomfortable to experience are often considered to have no redeeming value, to be avoided at all cost. Yet, perceptions of shame are not monolithic; being shameless often implies in the United States that one is ethically flawed or limited, as one is not able to experience one's own moral limits (Mason 2010). This schema implies that a capacity to feel shame is beneficial. Indeed, people in many cultures at many times have viewed the experience of guilt and or shame as a temporary negative in service of a larger positive end. For instance, Confucian thought generally holds that "the *capacity* for feeling" [emphasis in original] shame is a prerequisite to honor and virtue that should be cultivated whereas the emotion itself is one to be avoided (Confucius 1992, 127; Katchadourian 2010, 249, 252–3; Mencius 2009, 2A6, 6A6, 7A6).

Cultural perceptions of guilt and/or shame can shape the reactions of people experiencing them. For instance, a study of salespeople who experienced shame in front of customers, found that Dutch and Filipino clerks felt a similar degree of emotion but reacted quite differently. The Filipino clerks were more courteous and likely to work to repair the perceived broken relationship with the customer.

Table 3.7. Internal and external types of guilt and shame

Agent judging or feeling	The entity being judged or object of the feelings	State of being		Experience of moral emotion	
		Guilty	**Shameful**	**Guilt**	**Shame**
Individual	Self	Individual guilt	Individual shame	Individual guilt feelings	Individual shame feelings
Individual	Another individual in the group or the group itself	Membership guilt	Membership shame	Membership guilt feelings	Membership shame feelings
Collective	The same collective	Collective guilt	Collective shame	Collective guilt feelings	Collective shame feelings
Any agent (individual, membership, collective)	Any other agent (individual, membership, collective)	Externally judged guilt	Externally judged shame	N/A or in membership above	N/A or in membership above

Table 3.8. Comprehensive typology of guilt and shame

Possible influence or catalyst	**Agent judging or feeling**	**The entity being judged or object of the feelings**	**State of being**		**Experience of moral emotion**	
			Guilty	**Shameful**	**Guilt**	**Shame**
Individuals, memberships, collectives, societal values, cultural priorities, etc.	Individual	Self	Individual guilt	Individual shame	Individual guilt feelings	Individual shame feelings
	Individual	Another individual in the group or the group itself	Membership guilt	Membership shame	Membership guilt feelings	Membership shame feelings
	Collective	Collective	Collective guilt	Collective shame	Collective guilt feelings	Collective shame feelings
	Any external agent (individual, membership, collective)	Any other agent (individual, membership, collective)	Externally judged guilt	Externally judged shame	N/A or in membership above	N/A or in membership above

The Dutch were more likely to react negatively and withdraw from the customer. Dutch clerks were also less likely to think creatively or use available resources to address the issue. They were less likely to resolve the conflict with the customer, let alone to the customer's satisfaction (Bagozzi et al. 2003; Goetz and Keltner 2007, 166). Here we see indicators that negative assessments of the self are not necessarily degrading to the self and society but may promote beneficial change (i.e., relationship building) (Ronson 2015) if, as in this case with Filipino clerks, the experience is accompanied by a cultural vision that makes such change seem possible.

By recognizing the cultural influences on feelings and judgments of guilt and shame of individuals and collectives, I emphasize that agents are not alone in their guilt and shame. This notion will play a significant role in my ethical assessment of attempts to induce guilt and shame in Chapter 7, and suggestions of what is necessary to respond to guilt and shame, regardless of how they come to exist, in Chapters 8 and 9.

With this expansion of the typology to include cultural influences on guilt and shame, the typology is complete, at least for purposes of this project. It now includes states and feelings of individuals, membership groups, and collectives; internally experienced and externally induced or judged guilt or shame; and cultural influences on the valence of and appropriate reactions to these experiences as illustrated in Table 3.8.

4
Philosophical Arguments for Individuals, Memberships, and Collectives in States of Guilt or Shame

The typology of guilt and shame developed in Chapter 3 is significantly based in social scientific research and historical observations of what people report experiencing in the face of atrocities. These perspectives do, however, leave several important questions unexplored. What does it mean to be in a state of guilt or shame? What do we imply about individuals and especially memberships and collectives if we claim they can be guilty or shameful? What do we imply about agency and the moral emotions to maintain collectives can experience guilt or shame? These questions are critical to holistically understand guilt and shame and to facilitate our ethical assessment in later chapters. Thus, this chapter examines philosophical arguments about identity, agency, responsibility, guilt, and shame. I argue that individuals, memberships, and collectives can all be guilty and shameful and that collective and individual guilt and shame do not reduce to each other.

To come to these conclusions, I must establish what it means to exist as a membership and collective, whether and how they can act and act with intent, and the ways these actions influence their identity. If collectives meet these criteria, then they can act in better or worse ways according to their own ideals. To the degree collective failures to live up to ideals are just failures of actions, they entail guilt; to the degree they reflect on the collectives' identity, they imply shame. To ground the notions of individual and membership guilt and shame, I employ the work of several scholars each of whom contributes important elements to philosophically substantiate a part of the typology but whose contribution is insufficient to support the whole: Karl Jaspers, Hannah Arendt, Joel Feinberg, Virginia Held, Larry May, and Tracy Lynn Isaacs.[1] To develop my argument that collectives can have identity, agency, and a form of intent that is more than the sum of their constituent parts, I extend the work of May and Isaacs. I maintain that such collectives are

[1] I do not focus on environmental ethics in this chapter because most environmental ethicists, whether religious or philosophical, are individualistic in their analysis of agency and responsibility or are vague enough about how individual, group, and collective identity, agency, and responsibility may be related that they do not help explain how collective guilt and shame can make sense philosophically.

Environmental Guilt and Shame: Signals of Individual and Collective Responsibility and the Need for Ritual Responses. Sarah E. Fredericks, Oxford University Press (2021). © Sarah E. Fredericks. DOI: 10.1093/oso/9780198842699.003.0004

not just relatively well-defined collectives such as nation states, corporations, or social groups but also diffuse collectives which may contribute to climate change alongside individuals, memberships, and well-defined collectives.

4.1 Individual and Membership States of Responsibility

In the wake of the Holocaust, an increasing number of laypeople and scholars questioned whether individualistic notions of identity, agency, and responsibility popular in modern Western philosophy adequately accounted for the atrocities of World War II. Were individual people who claimed to be "just carrying out their orders" responsible? Did the German people as a whole bear responsibility for the atrocities of the war? Were individuals responsible for the actions of other individuals in their nation or for the actions of their state? Were individuals tainted by the actions of others? Was the nation tainted by the actions of its citizens or government? As Jaspers' work aims to answer these questions, it is a logical place to start when seeking a philosophical justification for the multidimensional typology of shame set out in Chapter 3. His willingness to admit and explore the importance of feelings and the existential and metaphysical dimension of human existence set him apart from many if not most twentieth- and twenty-first-century scholars of collective agency, responsibility, guilt, and shame and make him a particular promising interlocutor. Yet, as we shall see, Jaspers' work and that of others who examine group responsibility after him is ultimately insufficient to ground the typology because he does not sufficiently consider (1) the existence and significance of collectives or (2) the ramifications of feeling guilt and shame.

In *The Question of German Guilt*, Karl Jaspers aimed to address the above questions by differentiating four types of guilt (criminal, political, moral, and metaphysical). He distinguished criminal, moral, and metaphysical guilt based on whether courts of law, one's conscience, or God respectively judges one guilty based on one's action, motives, or being respectively (Jaspers 2001, 25–6, 65). His schema was intended to provide intellectual and practical clarity on matters of responsibility, guilt, and blame while acknowledging the feelings of guilt that people can experience in the wake of communal horrors. While Jaspers' study of metaphysical guilt indicates something about the shared existence of humanity, it, like criminal and moral guilt, still only operates at the level of individual rather than collective agents.

In contrast, Jaspers' political guilt aligns with my notion of membership guilt (though he uses the term "collective"). His notion helps ground this part of the typology and clarifies my thinking about whether collectives can be in a state of guilt or shame. Political guilt occurs when one is guilty in the sense of being responsible for things done by the nation one is a part of by virtue of the fact that one implicitly or explicitly participates in, benefits from, and can shape the system. Thus, the citizen is in some way responsible for his or her nation going to war because he or she is presumed to share in its benefits and, even if he or she didnot directly support the war through lobbying, voting, etc., he or she has the possibility of shaping the system in which the war occurs, say, by demanding an accounting of the war dead. As in criminal guilt, the judge of political guilt is external to the guilty party (Jaspers 2001, 25). The individual is judged, but the acts committed are acts of the state. Thus, Germans alive during World War II are politically guilty for the actions of the Third Reich, because they supported them directly, did not oppose them, or did not oppose them strongly enough to make a difference.

Jaspers however, carefully argues that the existence of political guilt does not mean that all citizens are equally guilty—some will have directly contributed to the atrocities or will have been in a role that gave them particular abilities to intervene in the political system. Yet, all German citizens have guilt (Jaspers 2001, 55). This guilt "is necessarily collective as the political liability of nationals [...] Yet this liability as such leaves the soul untouched" (Jaspers 2001, 55–6). In other words, his political guilt does not encompass my shame because in political guilt one's identity is untouched. Note also that political guilt is the result of the aggregation of individuals' actions (voting or not voting, etc.) because of their particular role as citizens rather than emerging from a collective agent. In other words, it is membership, not collective guilt, in my terminology.

Jaspers was not willing to admit that collectives as collectives can have responsibility, guilt, and shame because he recognized that labeling a whole people as fundamentally flawed led to harsh reparations after World War I, helped precipitate World War II, and enabled prejudices that fed the Holocaust (Jaspers 2001, 35–6, 41–2). To avoid such repercussions, he focuses on the individual even though his category of political and, in a lesser way, metaphysical guilt acknowledged some of the ways individuals are shaped by their groups. While I understand his reticence about collective responsibility given his historical situation and the risks of shaming collectives that are perennially dangerous, I am concerned with a different dimension of shame—how shame about individuals and collectives can enable anthropogenic environmental degradation by promoting inaction, or business as usual. Thus, I cannot just argue against or ignore collective guilt or shame because they can be dangerous.

While Jaspers identifies multiple key aspects of guilt including different judges, states, and feelings, of individuals and memberships, his work is not enough to ground the typology, as he does not sufficiently deal with collectives, consider membership groups beyond the nation state, or deal with shame. Several strands of scholarship after Japers took steps toward addressing these issues.

For instance, Joel Feinberg articulates three conditions necessary for a person to be responsible for a harm: (1) "that the responsible individual did the harmful thing in question, or at least that his action or omission made a substantial causal contribution to it," (2) that "the causally contributory conduct must have been in some way *faulty*," and (3) that there must be a direct causal link "between the faulty aspect of his conduct and the [harmful] outcome" (Feinberg 1991, 53). Virginia Held narrows the discussion by examining not a state of being responsible but the conditions necessary to hold someone accountable. According to Held, to be responsible, an individual needs to have been aware of what he or she is doing and needs to be able to "have done something else." To be morally responsible, the person also needs to "be aware of [...] the *moral nature* of the action [...] either the moral import of the kind of action of which it is an instance or as the moral value of the consequences it may produce" (Held 1991, 90). Held argues that these conditions do not require that the person knows every aspect of the action but do require that he knows "the kind of action he is performing" (90). By this she means that the person needs to know if in general the kind of action is one that can cause harm (i.e. throwing a bomb) and whether this harm is morally problematic.

If these requirements are necessary for responsibility, i.e., the status of being guilty, then the prerequisites to these conditions (having the capacity to act, the capacity to rationally assess situations, and the capacity of awareness and intent) are also necessary. Thus, many scholars have focused their attention on whether groups can have these characteristics, particularly agency and intent. Moving beyond nation states with their clearly defined boundaries of membership, decision-making structures, and identity for members, scholars have studied groups with more transitory boundaries and shifting populations, such as corporations (French 1979; Velasquez 1991), social clubs (Cooper 1991; Feinberg 1991), or racial or ethnic groups (Appiah 1991; May 1996; McGary 1991). In corporations and social clubs, the individual responsibility one has for the actions of the group is not only based on being a member as it is particularly shaped by the particular role one has. A CEO or president will have more responsibility for overall decisions than an entry-level worker or member. While one may have a specific responsibility because of one's role, one is also responsible for the actions one carried out, ordered, or approved in one's role, or for contributing to or failing to stop the group from doing those actions—possibilities shaped by that role. Also subsumed under this heading is the aggregate guilt of individuals in a particular group, i.e., the way each member of a golf club is responsible for the water, herbicide, and pesticide use of the course. The responsibility is just the sum of individual people's responsibility given their particular roles. These kinds of responsibilities of individuals in relationship to groups are those associated with membership guilt.

In short, these notions of membership responsibility are limited with respect to the types of guilt (and shame) discussed in Chapter 2, because they presume a clearly defined hierarchy that may be acknowledged by group members or may function to delegate responsibility from individuals to groups. But such features are not always present in the groups to which guilt (and shame) is ascribed—collections of neighbors or parents at a particular school who feel and elicit guilt and shame feelings in others; people in the industrialized world; and even all humans. To understand whether these types of informal groups may be responsible, i.e., in a state of guilt, it will be helpful to examine Held's notion of a random collective.

Held introduces random collectives in order to explore whether a formal decision method for action is required to have responsibility. She examines a series of cases involving what she calls a "random collection" of individuals—those that are not explicitly or intentionally formed into a group with a decision-making structure. In her case studies, random collectives may find themselves on a sidewalk or street or in a subway or train car (Held 1991, 94, 96). In each case, the random collective encounters a moral quandary—should they help a man trapped in a building, having convulsions, or being beaten? If so, how? Does the answer depend on the knowledge and skills of each individual or of the group as a whole? Without getting into all of the details of Held's assessments of these cases, which are unnecessary for this work, we can turn to her main point: the decision-making structure in organized collectives in such dilemmas makes them responsible to act by having a boss designated to make decisions, rules of voting, or customs (Held 1991, 97). Random collectives of strangers on a sidewalk cannot then be held responsible for not acting or failing to act to directly combat the danger because they have no decision-making structure. A random collective of strangers can be held responsible for "not constituting itself into a group capable of deciding upon an action" and carrying it out if "it is obvious to the reasonable person that action rather than inaction by the collection is called for" (Held 1991, 97). Neither the random collective nor its parts are responsible for other kinds of acts because, according to Held, they cannot perform them without having a decision-making structure. Each individual in the random collective is responsible for the failure of the group to act to form an organized group and then act, because each could have prompted such organization "though none of these individuals is responsible for not taking the action that ought to have been taken by an organized group in these circumstances" (98) because that action is a group not individual act. As such, it is something that the individuals cannot do on their own. For organized collectives, however, the responsibility of the collective is more than the sum of the responsibility of individual members because of the contributions of the decision-making structure (97). Consequently, while Held's analysis, like that of many who look at non-national organized collectives, moves beyond Jaspers' political guilt insofar as it extends the type of group studied, it only deals with a narrow part of the typology.

To ground the other dimensions of the typology beyond the focus on individual, membership, and some collective guilt states, we need to look to analyses that go beyond narrowly defined collectives, beyond determining who is blameworthy, beyond actions. We need to investigate emotions and identity in order to account for shame. To do this we need to explore the metaphysics of individual and collective identity which will identify where agency, responsibility, and emotions are possible.

4.2 Identity, Metaphysical Guilt, and Elemental Shame

Given the complexities of the metaphysics of collective agency, responsibility, and emotions, it will be helpful to start with a simpler case given the habits of our thought, that of the individual, and then build to collectives. As we look to individuals, Jaspers' vision of metaphysical guilt is a good place to start because he neatly illustrates how metaphysical status shapes identity and our interactions with the world.

For Jaspers, metaphysical guilt is rooted in the fundamental metaphysical status of being a human. As such, it is more aligned with my notion of shame than of guilt. Metaphysical guilt has at least three characteristics. Living continually causes harm to other living creatures as we breathe them in, step on them, or kill them for food (Jaspers 1969, 54). Second, and most emphasized by Jaspers, is a "lack of absolute solidarity with the human being as such—an indelible claim beyond morally meaningful duty" (Jaspers 2001, 65). Here, solidarity seems to be a fundamental shared status among human beings, "the substance of their being" (Jaspers 2001, 26) that gives them a coherent purpose and existence. He maintains that solidarity between people should impel us to care for others, though he notes that in fact there is often a great divide between this demand and what we actually do. This "capacity to live only together or not at all," to live in support of one another, is not embodied among humans as a species or even in groups of "fellow citizens or even of smaller groups, but remains confined to the closest human ties," i.e., of close family or friends (26). The tension between the actual solidarity of humanity at a metaphysical level and our ability to live out what this solidarity should imply or entail about our responsibility to and care for others is a part of being human for Jaspers: "If human beings were able to free themselves from metaphysical guilt, they would be angels" (27). He writes, "There exists a solidarity among men as human beings that makes each co-responsible for every wrong and every injustice in the world, especially for crimes committed in his presence or with his knowledge. If I fail to do whatever I can to prevent them, I too am guilty" (26). Indeed, the tension between solidarity and our ability to instantiate it is often framed by Jaspers as a failure to act to prevent harm to others: "If I was present at the murder of others without risking my life to prevent it, I feel guilty in a way not adequately conceivable either legally, politically, or morally. That I live after such a thing has happened weighs upon me as indelible guilt" (26). The bifurcation between ideal and actuality is itself a third key part of human existence for Jaspers (27). He writes "This solidarity is violated by my presence or wrong at a crime. It is not enough that I cautiously risk my life to prevent it; if it happens, and if I was there, and if I survive where the other is killed, I know from a voice within myself: I am guilty of being still alive" (Jaspers 1969, 52–3; 1970, 216–17; 2001, 65).

The shocking quality of this claim in the context of Jaspers' analysis of who is responsible for the Holocaust lead some, including May (1992, 147–8), to read metaphysical guilt as involving the lack of preventative action rather than the conditions of solidarity or fundamental inability to live out this solidarity distinct from any particular action. But when one considers that Jaspers intends metaphysical guilt to be something different from the other three types of guilt, it becomes clear that metaphysical guilt does not stem from inaction or action (Jaspers 1969, 55). Rather, an action is important to metaphysical guilt only insofar as it may make people aware that they are subject to the metaphysical condition basic to human existence. Jaspers maintains that no one wills themselves into being or wills this destructive existence. An individual is guilty just "because of what he is" (54).

Many aspects of Jaspers' metaphysical guilt are quite compelling. It is difficult to argue with the claim that our existence compels the death of others. His description of the chasm between the demands of solidarity and what we actually do seems to capture human experience. Haven't we all placed selfish priorities, short-term desires, or sheer laziness above people we claim to care about? If we cannot even prioritize them consistently, it is reasonable to assume we do not prioritize all with whom we are in solidarity.

Jaspers' concept of metaphysical guilt also helps us explain what it means to be in a state of shame. Here, solidarity is a condition of existence and ideal that we do not live up to, hence we are shameful and may feel shame. As such, shame is rooted in a worldview about metaphysical finitude and human existence as such. While not all shame will be rooted in solidarity with humanity (it may also be grounded in connection to other living beings, all entities, responsibility to God, etc.), it will follow the same format of an ideal that humans are unable to live up to.

Yet Jaspers' analysis is, once again, very individualistic. He either examines human existence as such or what human existence (through solidarity) entails about the status and actions of an individual. While individuals have solidarity in common, this commonality does not extend to a shared identity in groups or collectives.

Arendt's work, particularly her notion of elemental shame, provides us with several helpful ways to expand Jaspers' work to ground the parts of the typology dealing with individual shame, but is ultimately still individualistic. Arendt looks beyond the particular identity of an agent doing a problematic action to the category of being human. For Arendt, elemental shame occurs when one is ashamed of being human because one recognizes that oneself, like everyone else linked by the solidarity of humanity, has the capacity for evil. She arrives at this conclusion after reflecting on the events of the Holocaust. She notes that many of its perpetrators were motivated to participate to provide for their family's day-to-day needs or to keep a pension (Arendt 1991, 281–2). Social conditions facilitated such decisions. As such, she argues that if things were different each of us could have been the one running the gas chambers, prioritizing our own self and/or family over others (280, 282). Each of us as a human has the capacity to participate in evil.

Arendt's work adds four important elements to the experience of being human articulated in Jaspers' work. First, for her, the metaphysical capacities of humanity can give rise to evil acts, which in turn shapes our identity. These acts are not just possible because of a basic capacity of humans for evil, i.e., a metaphysical possibility of failing to live up to ethical ideals, but also because, for Arendt, humans are not just beings who have ideas, think, or feel, but are also beings who act. Our action help make us who we are (Arendt 1998, xvii, 7–8, 22, 179, 181). Second, she recognizes that evil acts are facilitated by particular social conditions and pressures. Her discussion of the capacities of administrative life to influence individuals and the ability to get caught up in the system illustrates how society's systematic forces shape individuals, actions, and identity more than Jaspers' analysis (Arendt 1991, 277–8). Third, Arendt expands the description of how one recognizes the human situation, the identity of being a human. It is not just through feeling Jaspers' metaphysical guilt or recognition of broken or unrealized solidarity or through reflection on your actions, but also through the recognition of commonality with other people that commit atrocities. Thus, Arendt's notion of solidarity includes not only existence as such but also solidarity with people who act in particular ways.

Arendt's attention to action provides a tighter link between the experience of being human and the specifics of immoral acts, or moral guilt, a fourth extension of Jaspers' work. Since each of us shares in the capacity for evil, Arendt argues that we all need to be responsible for the actions of everyone (Arendt 1991, 282) because the evil could have been committed by any one of us. Those of us who directly contributed to the crime are different from those who did not (277), and yet this difference is not absolute. It is one of degree, not kind. As people recognize the shared possibility stemming from our conditions of being humans and as acting beings, Arendt argues that they should act to stop the evil of which humans are capable: "upon them and only them, who are filled with a genuine fear of the inescapable guilt of the human race, can there be any reliance when it comes to fighting fearlessly, uncompromisingly, everywhere against the incalculable evil that men are capable of bringing about" (283). Those who recognize that they too could have done such evil if things would have been just slightly different have a special onus to fight it. Here we see Arendt articulating an individual responsibility in the face of actions committed by other individuals given their social situation and basic humanity. Interestingly, while this responsibility is ultimately borne by all people, there is a particular responsibility on those who feel (fear) and recognize the conditions of being a human—the shame feelings of the objective elemental shame of being a human.[2] Action, emotions, identity, recognition, and reflection all play a role here.

Even though Arendt better acknowledges the relationship between individuals and communities than Jaspers, and together they provide a ground for concerns of individual human identity and shame, her focus is still on individuals. Individuals are presumed to be the only agents. To ground membership and collective guilt and especially shame, we need a better sense of the identity of groups and their relationship to individuals.

4.3 Membership States of Guilt and Shame

May takes significant steps toward a theory of what I call membership states of guilt and shame. According to May, for "virtually every person, individual identity is intimately connected with group membership" (May 1992, 160), such that the actions of individuals in groups to which one belongs as well as actions of the group as a whole shapes one's actions and beliefs (151–2). Culture arises out of individual actions and beliefs; yet it is more than the sum of individual actions and attitudes, as something new can come out of the combination of people in the group. Thus being a part of the group opens individuals to new possibilities and, indeed, often guides their concepts, beliefs, and actions in particular ways (May 1987, 9–30). Despite this mutual influence, May stresses that it is important that individuals do not perceive themselves as wholly determined by their group membership because then "significant options for moral and social change are cut off" (May 1992, 160). Thus, he emphasizes that the identity of the group and individual continually shape and reshape each other, but that individuals can and should intentionally play a role in this process.

In this context, May investigates whether an individual who "is a *member* of a group that is causally responsible for a major harm" is morally guilty even if the individual as individual did not definitively causally contribute to the harm (May 1992, 151–2). May dismisses this type of moral guilt as inappropriate because it leads to people being punished for what they were not directly responsible for (154–5). Instead, he articulates another vision of responsibility, paired with a separate process of moral taint, to explain the interaction of an individual and his or her group when moral harm has been done by the group or another member of it. Both responsibility and moral taint hinge on his notion of solidarity and the action, or lack thereof, of the individual. For May, individuals of "a shared

[2] Jaspers makes a similar claim about the recognition of metaphysical guilt, but ties this recognition to taking action less explicitly than Arendt.

culture identify themselves with the group and in a sense with one another," i.e., they have a sense of solidarity. "The solidarity of a community" can give rise to a number of feelings about the collective, including "a shared feeling of responsibility for what occurs within the group (153–4). Indeed, solidarity means that people may still feel identified with group actions even if they disagree with them and try to distance themselves from them (154–5).

This sense of responsibility is not merely or necessarily about direct, causal action that one has participated in and does not hinge on a particular role one has in the group (i.e., a leader or voter) which gives one extra responsibility. Rather, it extends to moral taint or when one assesses oneself as morally diminished (May 1992, 152). Feelings of moral taint can arise because of a judgment that the group did something wrong that, as in Arendt's analysis, one realizes that one could have also done, or because one did not intervene to stop the wrong (149). This experience of moral taint or shame (May equates the two) may be appropriate, according to May, since it is based on an assessment of who one is, including (1) associations, (2) the basic existence of group relations, and (3) what one has chosen to be through actions and inactions. Critically for May, tainting actions can occur when one could distance oneself from the associations and chooses not to, resigning oneself to the situation, or refusing to believe that things could be different (158–9). In other words, one judges oneself and community as so problematic that one does not see a possibility of change—one is overcome by the assessment of shame. Without this hope, it is harder to see, let alone be something else, to push back against the wrongs of one's community. Thus, moral taint is metaphysically and morally significant because it shapes the kind of person one becomes. May argues that as taint involves a diminishment of the self, resisting it requires changing who one is in the face of community harm and acting to prevent or diminish such harm to the extent we can. It is in this sense that we need to take responsibility for the community we are in (161–2). May does not, however, sufficiently recognize the ways in which we may not be able to fully take responsibility: our social situation and the conflicting priorities of our ethics may hinder our ability to live well (Tessman 2005).

May rightfully points out that moral taint is not as severe as moral guilt in terms of being directly culpable for a harm (May 1992, 161), yet argues that it "takes seriously our group affiliations and thereby sets the stage for taking seriously our shared responsibility for the great tragedies which our communities have perpetrated" (162). Here May emphasizes his vision of individual identity being shaped by and shaping the communities of which one is a part and which may undertake problematic actions as a group. Given this more integrated notion of identity, it is reasonable to have feelings of membership guilt and shame about what one's group has done because one's identity is implicated in the action and the judgment that the actor, whether another individual or the group at large (and thus oneself), is problematic. The actions of others implicate you not just because of your actions or lack thereof, nor just because of your role or your recognition that you too could have done such a thing, but because your own identity is shaped by their existence and action. Your identity is eroded, you are thus in a state of shame. May maintains that if people think or feel they are morally tainted and act as if they are, then for all intents and purposes they are. He recognizes but does not strictly divide feelings of moral taint from a metaphysical status of being tainted. Thus, he provides a ground for membership shame—the shame of individuals about the group or other individuals in the group.

Despite these advances, May's approach is still limited because even though the individual and community are related in his work, his focus reverts to the individual when he discusses feeling and responsibility even if it is the individual in a social group. While this prioritization of individuals makes sense given the history of modern philosophy's individualistic bent, it is not sufficient to deal with all of the layers of responsibility, guilt, and shame that can and do exist. For that, we need a deeper analysis of what it means to be a collective. The work of Isaacs, combined with other aspects of May's work, will get us one step closer to that goal.

4.4 Collective States of Guilt and Shame

For Isaacs, individuals and collectives can both be responsible for acts arising out of their respective individual or collective agency and intent (Isaacs 2011, 12, 52–3). Moral responsibility is "the blameworthiness or praiseworthiness that we attribute to agents for their actions. Agents warrant praise when they do the right thing" (53). To have moral responsibility is to be guilty in the sense of culpability or of objectively committing the act. Isaacs asserts that there are two types of collectives that can have collective responsibility, i.e., guilt: organizations and goal-oriented collectives. Organizations have a formal structure and generally have clear rules of decision-making. They often have a system of leadership such that some people are more responsible than others for the actions of the group, yet the collective still has actions and intentions that are more than the sum of the actions and intentions of individuals in the group (24, 28). Common examples of organizations include "corporations, nonprofit groups, nations, universities, departments of philosophy, and professional sports teams" (24–5). Goal-oriented collectives are, as their name implies, collectives that are defined by a shared goal that they act toward with collective agency (25, 32). Isaacs' prime example of a goal-oriented collective is the collective that perpetrated the Rwandan genocide, though she also notes that "two people going for a walk together, three people painting a house together, a thousand people doing the wave at a sports event, or tens of thousands of people 'Running for the Cure' all constitute goal-oriented collectives" (25).[3]

Agency, whether individual or collective, hinges on the capacity of intentional action (Isaacs 2011, 27, 53). In collectives, both intentionality and action arise from "a collective intentional structure" (27). This structure in an organization, such as the Canadian Red Cross, Isaacs' example, includes explicit "authority structure[s], [...] policy [...] procedures for making organizational decisions and for the organization taking action" (28–9). Goal-oriented collectives are less structured than organizations, but their ability to articulate and desire goals that an individual would not prioritize as an individual and their ability to persist as members come and go, among other features, suggest to Isaacs that they have identities that are not merely the sum of the parts of their members (31–6). Beyond identity, however, the goal itself is critical to Isaacs' understanding that goal-oriented collectives have intention and agency and therefore responsibility. She maintains that "when we intend something collectively, there is an understanding that we, as a unity, intend to bring it about together" (37–8). The goal itself binds the collective and demonstrates the collective's intention and agency. For Isaacs, this intention is an "individual intention with collective context" that is part of the collective intention which we intend to do in a coordinated way (40).[4]

Isaacs requires a number of conditions in order for a collective to have an intention. While employing an example about a two-person collective, Isaacs maintains that such conditions can and do apply to much larger groups as well: "each participant in the collective act has the appropriate understanding of the quite specific collective goal, her respective contributions toward the goal, her intention as contributor to contribute, the intention of her coparticipant[s] to do the same. Under these conditions, they pursue the goal as a unit" (Isaacs 2011, 36, 46). Elsewhere Isaacs writes, "Each needs to be *aware of* the collective goal as a collective goal if the action toward its intention is to be a collective action" (45; emphasis mine). Note that the knowledge of the situation, self, and group is necessary and that this knowledge needs to be explicitly acknowledged. The combination of elements yields an explicitly acknowledged group acting intentionally. In sum, Isaacs finds that responsibility can apply to both organizations and goal-oriented collectives because they can have two criteria for responsibility: recognition of the goal and the ability to act with intention.

[3] Isaacs also delineates two other even less organized types of collectives but maintains that neither has responsibility because she thinks they do not have agency. Thus, I do not examine them in detail in the context of her work, though my analysis of responsibility for climate change counters her claims that these other groups are incapable of responsibility.

[4] She sees a continuum of collectivity as she recognizes that the goal-oriented collective is not as tightly knit as the organization.

While Isaacs helpfully parses the difference between organizations and goal-oriented collectives, her emphasis on intent—the need for group members to explicitly recognize participating in the group—means that her work does not sufficiently deal with anthropogenic environmental degradation in general and climate change in particular, the most complex and challenging case of responsibility, guilt, and shame. In Isaacs' analysis, climate change is just the result of parallel actions of many individuals (Isaacs 2011, 45, 48, 100). There is no *collective* responsibility for climate change because no goal-oriented collective, let alone organization, causes it or aims to cause the change (153). The changes were not even a goal that united a previously disparate group into a goal-oriented collective. Without the agency and intent that comes with these collectives there is no collective responsibility for climate change, and it is an unintended result of activities done by individuals (130). Each individual is only responsible for his or her small individual contributions to climate change (26).[5]

This account of responsibility is insufficient because it does not recognize the ways that collectives and systems condition individual contributions to climate change and how collective contributions to climate change are more than the sum of the parts of individual actions. We need to be willing to rethink the conditions of responsibility in the face of climate change to meet its challenges. In particular, I propose that conscious intentional acts of explicitly acknowledged collectives are not the only acts that deserve moral consideration. Unintentional acts of unacknowledged collectives, such as the greenhouse gas emissions from the diffuse collective of industrialized people, that function as goal-oriented collectives also indicate a form of moral responsibility.

It is possible, however, that some goal-oriented collectives and their intentions may only be identified after the fact rather than explicitly and prior to action, as in Isaacs' work. May suggests just this with his notion of prereflective intent developed through an analysis of the solidarity and collective action of mobs. For May, solidarity is a cohesive relationship between individuals (May 1987, 40, 62) that enables a shared purpose to arise. His prime example is the Paris mob that stormed the Bastille in 1789. Like Isaacs' goal-oriented collective, May argues that the participants in the mob did not have an explicit shared decision-making structure (58–9). They had no rules or mission statement, no policies or procedures, no leader giving orders or coordinating the effort as a whole. Unlike Isaacs' goal-oriented collectives, the mob members do not have an explicit identification with each other as being a part of the collective (294). Yet, over several days the members of the mob generated ideas and coordinated their efforts to do things

[5] Isaacs argues that there needs to be collective forward-looking action about climate change motivated by obligation rather than responsibility. She argues that obligations are incumbent on collectives, individuals, and as yet unformed groups that could do something to solve a problem (Isaacs 2014; 2011, 130–55).

that none could have done alone. Something new happened in and out of the mob. Drawing on the work of Sartre and psychological research, May attributes the collective action of the mob to prereflective intentions arising from the mob's solidarity. Such intentions are not explicitly articulated by an individual prior to the mob's emergence or even during this time. Rather, they emerge out of the mob (61, 63–4). May argues that because these intentions are beyond what the individuals would intend on their own, they can "in a limited way, be ascribed to the group" (64). Therefore, though the individual members of the mob are not aware of the interactions, the mob "may be treated" as if it has collective intent and purpose because it has intents and actions, and these are different from what the individual would have on their own. May is careful to say that the mob does not intend in the same way that individuals do, and therefore there are not "collective intentions proper" (64). "Rather each member of the group comes to have the same intention, either reflectively or pre-reflectively, and this is different from what their individual intentions would be if they were not members of the group. The sameness of intention is collective in the sense that it is caused by the group structure, that is, it is group-based" (65).

If we accept May's account of intent, and that collectives can act as collectives, then it also makes sense to attribute responsibility in some way to the group as such and not just to the individual. I, however, go further than May because I reject the idea that mob intentions are not "proper intentions." Rather, mob intentions are a particular kind of intention, those of inexplicit collectives.[6] If the people involved in a prereflective mob are not aware of the intent of the mob, then they also are not necessarily aware of being in the mob, or even that there is such a group that could be delineated. Yet, people may act in ways that functions as a goal-oriented collective and deserves the attribution of responsibility retroactively. A prime example of this occurs in climate change. For example, consider as a group all who participate in development through industrialization and high rates of consumerism. This amorphous group contributing to anthropogenic climate change is made up of many individuals as well as subgroups including what Isaacs calls organizations (fossil fuel corporations, car manufacturers, agribusinesses, nations with subsidies for fossil fuels, etc.) and goal-oriented collectives (e.g., residents of a town who desire clean water, reliable electricity, or telecommunications).[7] This conglomerate is a large unconscious, goal-oriented collective with intentions and actions, what I name a "supercollective." Its main goal includes industrial development through high rates of production and consumption; it typically views humans as radically distinct from (and superior to) the rest

[6] I take this position because I do not want to ontologically privilege individual intent even if reflection on individual intent informs my understanding of collective.

[7] Others also contribute to climate change, but this group has done the most and is the one I am primarily concerned with.

of nature. Its actions, whether facilitating trade in economic policy, building transportation infrastructure, or developing and using more disposable products, contribute to these goals.

Such actions and the belief that the actions are preferable are not merely carried out by individuals in parallel with others, as Isaacs would claim. The existence, agency, or responsibility of a supercollective is not reducible to collectives or individuals that are a part of it, as the existence, agency, or responsibility of a collective is not reducible to the individuals that are associated with it. The actions and beliefs are dependent on and interwoven with the physical conditions of the world, including laws of physics; chemical, geological, biological, and evolutionary conditions; the physical infrastructure of human civilization; and the actions, beliefs, and social structures of others in the past, present, and, to some degree, future. The collective would not exist as it does without fossil fuels; access to technology to extract, refine, and use fossil fuels; subsidies for these technologies and their implementation (e.g., subsidies for highways and wars to protect our access to resources abroad); dreams of "autonomy" (enabled by but predicting the automobile); the extreme prioritization of economic value; and, as Michael Northcott persuasively argues, a worldview predicated on the separation of humanity from the rest of nature, including the presumption that humans are unable to significantly affect the environment including the weather (Northcott 2013). These conditions shape what the collective enmeshed in them conceives as possible, how it feels, and how it acts. Admittedly, most participants in this group would not name it as an explicit part of their identity. They may be more likely to name themselves with respect to gender, race, nationality, religion, clubs, alma mater, sports teams they support, etc., than to say they are industrialized or participants in a capitalized society. Yet, this identity so deeply informs many in the contemporary United States that they cannot imagine existing otherwise. Acknowledging it is like a fish acknowledging its water. How else could it live?

In this situation, it is reasonable to drive by oneself across town to pick up a gallon of milk or even an unusual ingredient for a new recipe. It is reasonable to fly across the country for a few days for work, to see family, or just for pleasure. It is reasonable to ship in meat or fresh flowers from around the globe. It is reasonable to give fossil fuels huge subsidies, to deforest the rainforest, to have broken the prairie. Indeed, these things are so reasonable that they are not generally considered, let alone judged as problematic. Insofar as these assumptions shape how individuals, collectives, and supercollectives act and conceive of themselves, these assumptions also shape identity. Contrary to Isaacs' analysis, goal-oriented collectives do not have to be explicit about their group identity or goals to exist as a collective.

Existence is not the only issue at stake; agency and responsibility are as well. Again, May's work is a helpful starting point. He analyses the degree to which individuals are responsible for holding attitudes that bring about harm in society even if the individual does not directly cause the harm. May rightfully argues that individuals are responsible for holding attitudes that increase the likelihood that others will do harm if and when they "create a climate of attitudes in which harm is more likely to occur" (May 1992, 46). This can occur if people with the attitudes do not take precautions against harm or if they enable social structures including policy or traditions that enable harm. While his examples are racism and sexism, the form of his argument also works with strong anthropocentrism, environmental racism, or climate denial.

While I agree with May's analysis of the degree to which the individual is responsible for their own attitudes in relationship to the group, once again, I want to push his ideas further. The supercollective as supercollective or collective as collective is also responsible, because it enables and sustains attitudes, institutions, structural inequalities, physical infrastructure, etc., which foster and reinforce the worldview and acts that lead to climate change. In other words, the unintentional climate change-producing attitudes, structures, and acts of unacknowledged supercollectives such as industrialized nations and peoples enable them to function as goal-oriented collectives. Therefore, they have moral responsibility alongside that of organizations such as nations. We who benefit from past activities leading to climate change also have inherited collective responsibilities for the disproportionate burdens our collective has placed on others.

Once we have existence, agency, and responsibility of collectives, it is reasonable to ask whether collective responsibility supplants individual responsibility or exists alongside it. Isaacs' two-level theory of responsibility in which individuals and collectives are responsible for acts arising out of their respective agency and intent is a helpful model (Isaacs 2011, 12, 52–3). Critically, for Isaacs, the individual level is neither subsumed nor replaceable by the collective and the collective level is not reducible to the individual even though each can influence each other's attitudes and actions The two kinds of responsibility are separable (Isaacs 2011, 24, 52–70; 2013, 46).

The case of climate change indicates that two expansions to this theory are needed. First, multiple collectives can each be partly responsible for an event. Second, as alluded to above, collectives may be a part of one or more supercollectives, which are related to collectives in a similar manner to the way collectives relate to individuals. Individuals relate to supercollectives as individuals or through collectives. Thus, multiple layers of responsibility for different aspects of a complex event like climate change may exist simultaneously. To the degree a collective has responsibility and ideals, it also has the possibility of failing to live up to its ideals and thus be in a state of guilt.

To ground the notion of the state of collective shame, we need to move beyond Isaacs' work, as it does not sufficiently consider the co-constituting relationships of individuals and groups or the metaphysics of groups and therefore does not deal with shame. Recall that shame stems from an entity being less than its ideal.

With respect to individual humans, there are many ways to conceive of this—of finitude in general, of the failure to live in solidarity with other humans, etc. Memberships then can have shame due to the shame of their component parts and the sum of such shame. Collectives, however, are more than the sum of their components, they have supervening characteristics. Thus, the issue is whether one of these supervening characteristics is the ability to be perfect, to be congruent with the ideals of a collective and, thus, without shame. I do not think so. I have seen no empirical evidence for the perfection of collectives (whatever that may be). Furthermore, claiming that it is the case would require additional metaphysical assumptions about collectives that I am not prepared to make because I see no experiential or philosophical support for such assumptions. Why would collectives be different from anything else? Therefore, I assume that collectives are inherently limited. Insofar as the ideals of collectives or supercollectives may outstrip their capabilities, collectives and supercollectives have the possibility of being in a state of shame. As with individuals, I am reluctant to say that being shameful completely describes the entity, for it seems that many agents have both shameful and positive characteristics at their core, and that a particular characteristic may be positive or negative in different situation. The state of being shameful may be one of several characteristic features of the collective.

4.5 Experiences of Collective Guilt and Shame

Having provided an interdisciplinary grounding for the possibility of collective emotions, especially guilt and shame, I will take a moment to summarize what collective feelings entail. Collective feelings are experienced by the collective about the collective, or experiences by a subcollective of the collective about the whole, as when some from the United States feel guilt and shame about their nation's environmental policy or lack thereof. Given the prioritization of individuals in our culture, we may not perceive collective emotions as collective even when they exist. We can know them to exist by their subject and shared character. They are emergent properties of the group and can spread throughout it. These emotions are justified by the existence of states of collective shame, as discussed in earlier sections. Indeed, states and feelings are mutually influential.

Admittedly, membership guilt and shame may be hard to differentiate from those of collectives. This is because the ontological difference between individuals and memberships and between memberships and collectives is not absolutely distinct. Identity is less quantized and more interconnected than we often realize, and yet both individuals and collectives have some agency of their own, while membership responsibility is the sum of their individuals' responsibility. Consequently, memberships do not have as much ontological distinctiveness as collectives, since their responsibility comprises the responsibility of their parts, each in their own role. One could capture all of membership responsibility, guilt, or shame by looking at the characteristics of the members. While the category of collective makes more philosophical sense than membership, sometimes it can be pragmatically useful to use membership instead of collective. This can happen when you want to emphasize the roles involved and how guilt and/or shame can occur because of them. It can also happen when you want to focus on particular actions and the small-scale influences that led to them rather than the whole collective system. Acknowledging the collective is important for recognizing the emergent responsibility and feelings of collectives and for determining how to respond to them. It is also critical to recognizing the different agents simultaneously responsible for an outcome such as climate change

If one is keen on differentiating between membership and collective guilt or shame, a couple of rules of thumb can be used. An agent using "I" over "we" language with little to no recognition of the group or social forces is the hallmark of narratives of individual guilt or shame. The scale of analysis may also help one identify the differences—when the whole group versus particular individuals in the group are considered. The focus of analysis—whether something wrong done by someone in a particular role or done of and by the whole system—may also be helpful. Finally, one can look to the identity in question, whether a person or collective as a whole.

Admittedly, I have not definitively proven that collective emotions exist. Given the nature of knowledge and proof, such absolutes are not possible. The best one can do is to (1) indicate how other theories, e.g., individualism and extreme rationalism, are not able to explain experience or suggest an ethical response, and (2) show how a number of types of evidence corroborate one's own theory. The wide variety of evidence supporting this idea of collective guilt and shame, both states and feelings, allows us to articulate a fuller picture of what it means to be a part of a collective and, indeed, to exist. The recognition of collective feelings enables a tight coupling between our metaphysical and moral experience at all levels. Aside from the intellectual satisfaction of a more comprehensive, coherent theory, this typology yields practical benefits as well. For instance, paying attention to moral emotions can help us recognize when and how new morals are emerging in theory and/or through ethical practice. Especially when this happens at a collective level, societal and policy and infrastructure change may well be not far behind. Recognizing such collective emotions helps us understand societal phenomena, and thus better develop ethical theories, effective techniques of moral persuasion, and responses to affective experiences, which are subjects of later chapters.

Now that I have articulated foundations for the typology, I can return to the evidence of environmental guilt and shame to ensure that it meshes with the typology. This will also help us better understand the phenomena and determine how to respond.

5
Environmental Guilt and Shame

With a typology of guilt and shame at hand and arguments about why individual and collective states and feelings of guilt and shame are justified, we can now return to the evidence of environmental guilt and shame to determine which types are experienced by contemporary U.S. Americans. After all, taking the experience of everyday environmentalists seriously requires testing to be sure the theoretical apparatus used to describe such experience aligns well with that experience. Additionally, it is also appropriate to test categories to ensure that they fit the data.

This chapter considers individual, membership, and collective guilt and shame in turn, followed by an examination of the induction of guilt and shame by outsiders (to the extent these overlapping phenomena can be differentiated). I will show that while individual environmental guilt and shame are most obviously judged and felt, there is also strong evidence of membership and collective guilt and shame states and membership feelings, as well as significant indicators of collective guilt and shame feelings in the English-speaking developed world, particularly in the United States. As collective guilt and shame are likely to be most controversial to readers and have the least evidence in the previously examined data, I will supplement the evidence of Chapter 2 with additional popular and academic literature.

5.1 Individual Guilt and Shame

Individualistic perspectives frequently dominate popular and academic literature about environmental guilt and shame. Some texts have a radically individualistic perspective, where the author is the only agent recognized; others acknowledge or desire support from other people but are still fundamentally individualistic. Let us first look to signals of extreme individualism.

Most obvious is extensive, often exclusive, use of the personal pronouns "I" or "me" (Crislip 2012; Gwen 2009a, 2009b; Helmer 2005; Hickman 2005, 146–7; Joy 2012). For example, in an article about the effects of a new San Jose law requiring merchants to charge ten cents for a paper bag, Robyn Roark admits her habit of forgetting her reusable bags and her resulting guilt: "Confession: I have no less than 1 million reusable shopping bags. I've had them for years. With every blogging event, I get a handful more... I always felt a little guilty when I forget my

Environmental Guilt and Shame: Signals of Individual and Collective Responsibility and the Need for Ritual Responses.
Sarah E. Fredericks, Oxford University Press (2021). © Sarah E. Fredericks. DOI: 10.1093/oso/9780198842699.003.0005

reusable bags" (Roark 2012). The use of "I" or "me" in every, or almost every, sentence is common in these narratives. Jodi Helmer's blog post, first discussed in Chapter 2, in which she reflects on her "eco-sins" as she washes the dishes, uses "I" or "me" nine times in eight sentences. She is the only actor she explicitly considers, and implies that she is the actor behind events with no directly attributed causes such as the open back door letting the air conditioned air out (Helmer 2005). Helmer's self-focus includes both her outward actions (washing dishes, driving, throwing things away) and her inner thought processes (thinking, obsessing).

Many of these narratives portray being a good environmentalist as a solo affair. It is up to the individual to live life correctly and overcome past infractions. Jennifer Willis's "New Year's resolutions are about hope and change and moving forward, and *it's up to me to wipe my own slate clean*" (Willis 2008; emphasis mine). She takes responsibility for her own actions and for all of the change that must occur. She aims to set her own goals, determine whether or not she can do better, and motivate herself. She does not acknowledge the possibility of others serving to keep her accountable or providing her moral support (and notes that her boyfriend "laughs at my green guilt") (Willis 2008). Willis is out to change her environmental behavior on her own. Here and elsewhere, authors not only emphasize individual actions leading to individual feelings but also convey that the proper response is a solitary activity (Hickman 2005, foreword, 1).

Indeed, authors may take an individualistic perspective even when the actions in question are necessarily communal. Consider a comic by Dan Piraro in which a man in the woods holding a chainsaw connected to a solar panel on his back remarks to his companion "This solar powered chainsaw makes me feel less guilty about deforestation" (Piraro 2012, 9). As he emphasizes his personal guilt feelings, he fails to recognize the communal nature of deforestation. This term is typically reserved for large-scale clearing, either through the combined effect of many small actions or through massive corporate timber operations. An individual would not have much impact or reason to cut so many trees without communal desires for timber or cleared land. The solar-powered chainsaw user tries to cope with a large, relational issue as an individual, the absurdity of which is part of the joke. Similarly, blogger Gwen places all the blame for giving *and receiving* too many Christmas gifts, "a nauseating display of consumption and waste" on herself, not recognizing that two parties are necessary for gift-giving or that social expectations encourage such behavior (Gwen 2009a).

Interestingly, emphasizing the individual often persists in environmental self-help books in which authors are explicit about sharing a home with others or being in relationships (e.g., with friends or coworkers). For instance, though Robyn Harding has a husband and children who enter her narrative from time to time, and her children are the ostensible reason for the book, the decisions about their household environmentalism are almost always portrayed as hers and hers alone. This perspective starts in her title: *Mom, Will This Chicken Give Me Man Boobs?: My Confused, Guilt-Ridden and Stressful Struggle to Raise a Green Family* (emphasis mine) and continues through most of the book. Harding is the main actor. *She* learns about and buys some local and organic food, *she* reflects on the changing environmental rules as they move from place to place, and *she* feels bad when they cannot recycle any more or when she perceives her new neighbors as judging her lifestyle (Harding 2009, 46–67). There are few signs that her family members share these concerns or should have responsibility to act greener on their own. The problem—whether a bad action or uncomfortable emotion she wants to avoid—is with her. While one can imagine a text in which family members or friends reflect together on the process of greening their lives, or in which the author includes the perspective of his or her family members, most environmental books and blogs have a singular, individualistic voice.

Most psychological and market research about environmental guilt and/or shame also takes an individualistic approach to these topics. For instance, participants are frequently asked the degree to which they agree with individualistic statements such as "I feel guilty about the impact I have on the environment" (Lexmark International 2009; National Geographic and GlobeScan 2012a, 77; Qi and Roe 2016). Since individuals are the only kind of agent examined in the vast majority of such studies, results which focus on individuals should not be surprising nor taken as evidence that collective guilt and shame do not exist.

While the individualistic point of view is often the most readily apparent, it is not the only approach taken to environmental guilt and shame. Even those who are consistently and predominantly individualistic often acknowledge the need for community support, particularly regarding information. Authors seek ideas about being a better environmentalist from their blog readers (Gwen 2009a). Hickman asked environmental professionals, to come to his house to evaluate his household's environmental impact and give tips on how to do better. He enlisted their help because he desired accountability, "someone to kick my weak-willed, unethical butt whenever I wavered and started to show signs of giving up," and information "explain[ed] with reason—and preferably some hard science—[about] why I needed to make changes to my life" (Hickman 2005, introduction). Here we see attention both to changing concrete habits (i.e., using dishcloths instead of paper towels) and to himself—to develop a stronger will. This speaks to the desire to ameliorate guilt (about action) and shame (about negative aspects of one's identity). Often, admitting the need for information and moral support from others is the first step away from extreme individualism toward a relational perspective.

The move to recognizing and taking account of relationships is not easy; many struggle with conceiving of themselves not just as solitary actors but also as part of a larger unit, such as a household, family, or nation. Hickman, for instance, wants himself and his family to be greener. Tension arises because he is concerned about how his wife, Jane, is participating in and being affected by the project. He writes:

> I feel as though she is being dragged along, usually grudgingly [...] I really don't want it to be like that [...] I feel frustrated that she doesn't want to embrace things to the same degree as me [...] I never considered for a moment that one of the hardest aspects would be making it work and progress within a relationship. I'm at a loss what to do. The more I try to persuade her, the more she recoils. Yet I don't want to do all this without her by my side. Should I just leave her to work out her own course and determine her own level of engagement, even if it leaves me frustrated that we can't advance at the pace I want?
>
> (Hickman 2005, 125)

Here Hickman describes the disparate priorities of his family. He feels responsible for pushing the family toward environmentally friendly positions even as he realizes it is not fully up to him.

Indeed, people often compromise their environmental principles because of relationships, even though this means they incur more environmental guilt. As a commenter on a *New York Times* article remarks

> When you are in a relationship, it's especially hard to implement all of the environmental action-items you would ideally want. Your partner may agree with you about recycling and reducing waste, but canot [sic] stand the fact that you completely power down your computer, microwave, and entertainment center for 4/5 of the day. Not only do some of [us] deal with the guilt of being hypocrites by ourselves, but also with the guilt of having to compromise for the sake of a significant other. (Wadler 2010)

As with Hickman, the commenter focuses on himself or herself as an individual in a relationship.

Authors may also look beyond narrow individualism when they use tips, advice, inspiration, and shaming to influence others to do better. One writes

> I came to the conclusion after listening to and reading about these topics that I can only change my realm of influence. And though [sic] changing my sphere, others will be touched and will in turn make changes towards a green lifestyle.
>
> What I also realized was that I cannot, nor do I want to, take on industry. There are many groups who have been fighting unethical practices in big business for years. That is their domain! My domain is the 900sq meters of this planet that I call home and how I live on it.

> My sphere of influence reaches from my husband and children, to extended family and friends, and today, to you, on this website! (Jones 2013b)

Many authors, whether of blogs or books, share this sentiment. They aim to catalyze environmentalism through their prose and household activities. Yet, many still largely belong on the individualistic side of the spectrum. They may inspire or take inspiration or advice from another, but ultimately each person is presumed to be acting on his or her own, being guilty or shameful and feeling guilt or shame as an individual.

Multiple forces may contribute to this individualism. The memoir style of the environmental narratives may skew the literature, as may the modern Western culture in which the authors live with its predominantly individualistic worldview, value system, and language. For instance, marketing campaigns often emphasize the importance of individual consumer acts over changing industry or social systems through regulation.

The individualistic tone of women authors may also stem from a traditional gender dynamic where the wife is in charge of running the house and the husband is off at work, not wanting or allowed to be bothered with the daily minutiae. Indeed, this seems to be at least partly the case in Harding's work, as she rarely discusses her husband's input or involvement in her household's environmental activities. An author, regardless of gender, may also be much more invested in environmental values than others in the family, so sets off on his or her own (Willis 2008). As we shall see in Chapter 7, many of these forces promoting individual environmental guilt and especially shame mutually reinforce each other.

5.2 Membership Guilt and Shame

For some authors, responsibility, guilt, and shame stemming from a particular role segues into membership guilt. Recall that membership guilt or shame includes people feeling emotions or judging their status about things their fellow group members did. Pure examples of membership (as opposed to collective) guilt or shame can be difficult to find because the authors are not writing to elucidate these philosophical categories; the categories themselves may blur into each other; and the categorization of an event, state, or feeling may depend on the scale of analysis used by the categorizer. Yet, strong examples do exist.

One occurs in a *New York Times* interview about carbon offsets offered by airlines and travel companies. When asked why she participated in an offset program, "Martha L. Delaney, a lawyer in Minneapolis and a regular flier to San Diego" replied, "The fact that the U.S. government hasn't ratified the Kyoto treaty against global warming was a big reason for me" (Rijnen 2003). Delaney took action because of what governmental officials, members of her group, the United

States, had not done. She identifies a problem with their action and, since she is associated with them, wants to act to counteract their inaction so that the group as a whole is less problematic. She likely also *feels* guilt for what the others have not done.

I classify Delaney's action as membership rather than collective guilt for several reasons. Although she identifies as a U.S. American, she understands that the entity who has done wrong, the U.S. government, is separable from her. Additionally, she takes an individualistic response to this guilt—buying carbon offsets for what is likely her largest personal source of emissions, flying. Were she to engage in collective action (e.g., lobbying her representatives, organizing her community) or thinking of the problem in deeper terms of "we" rather than "me," she would be more on the collective side of the spectrum.

Membership guilt or shame can also happen when an individual recognizes that they participate in wrongdoing alongside the other group members. For instance, Jen Pleasants attempts to clarify the psychological responses to participating in environmental degradation and to validate that her feelings are real and significant:

> One [emotion] is actual guilt: when you feel bad because sometimes you knowingly don't do the '"right" green thing. This is different than eco-anxiety, which is when you feel overwhelmed and helpless because everything you and the rest of the world are doing feels as though it is having some kind of bad eco-repercussion. (Pleasants 2009, 11)

Pleasants' idea of eco-anxiety does not rest on being a solitary individual but rather on being one who is doing bad things alongside others. These parallel wrongs amplify the physical degradation of the environment and one's emotional response. Pleasants' account is about guilt because it concerns actions, and to the extent it concerns existence, the condition of being helpless to do otherwise, it involves shame. Her description starts to shade into the realm of collectives, but insofar as it is still about the summary of individual actions rather than assuming "we," it is still tinged with membership.

Indeed, many everyday environmentalists exhibit membership guilt or shame as they recognize that their individual actions have not caused the environmental problems and will not be enough to solve them (Kornblum 2008; Wadler 2010). They recognize themselves as members of a larger group, whether a citizen in a particular nation, a person in a developed country, or a human in general, that has done wrong and needs to act. They may see the group and themselves as fundamentally flawed. These examples are of membership rather than collective guilt or shame because the individual is still differentiating him- or herself from the collective by using "I/they" language—noting that he or she did not do the action in question, or is not alone in their action or status.

5.3 Collective Guilt and Shame

As people start talking about what "we" need to do, they move further toward the collective end of the spectrum. For example, von Jessen uses a number of "we" statements referring to industrial society after reciting a litany of facts about environmental degradation: "It seems like all news is bad news—and we are the culprits" (von Jessen 2009, 235). "Can we change our ways, allow nature to recuperate, and restore harmony between humankind and nature? I think it is possible" (239). Similarly, Vaze reflects upon the reactions of environmentalists as a whole when thinking about air travel in a time of climate change: "The thought of travel fills us with longing and guilt at the same time" (Vaze 2009, 287). This blurring of lines between membership and collective may be due to imprecise language or the lack of philosophical reflection on the part of the authors. Alternatively, it may signal a lack of strict ontological division between membership and collective guilt or shame. For all of these reasons, it can be difficult to determine when people see themselves and/or actually exist an individual unit in a group whose members can have an additive effect (membership) and when a coherent group exists with separate actions and identity from those of individuals. Yet, many examples indicate the existence of collective, not just membership guilt and/or shame.

Hickman identifies the Western, developed world as a culprit: "The irony of our Western lifestyle, of course, is that we are blissfully ignorant of the negative impact it has on ourselves, our neighbours and the environment, but that we choose to journey on regardless, blinded by a convenient fog of inertia and apathy." He maintains that the modern Western lifestyle is blind to "mercury-filled salmon, deodorants being linked to cancer, increasing levels of social exclusion and inequality, or landfill sites reaching full capacity" (Hickman 2005, 2 of forward). Similarly, von Jessen describes an exchange with a friend in which she remarked: "If we don't change our ways, we'll face a future of global hunger and starvation. I question whether we will survive as a species. Maybe nature will get her revenge—punish us for our exploitation" (von Jessen 2009, 160). The problem these authors identify does not relate just to an individual's actions but to the priorities, infrastructure, and collective identity of Western culture.

A collective perspective is even more obvious when people reflect on what it means to be human. In *No Impact Man*, Beavan often wrestles with whether he in particular and humanity in general is totally morally corrupt, given his and our impact on the planet. He tries to avoid the existential paralysis that can come with such a shameful assessment by blaming our environmental situation on the problems of the industrial system rather than a flaw in human nature (Beavan 2009, 20, 25, 40, 64, 65–6, 190). The fact that this feeling of whole-self-judging shame is a temptation for him, even if he ultimately does not think he should feel it, testifies to the fact that it is a concern of his milieu, especially as he is also interacting with his readers whom he hopes to educate, entertain, and inspire. When he is tempted by the whole-species-judging shame, his understanding of identity includes a collective identity, the vague group of contributors to climate change. People like Beavan (when he has not talked himself out of it) may identify with the group causing climate change enough that they absorb its identity whether the group is their nation (the United States in Beavan's case), the group of all industrialized people, or humanity as a whole. This identification may be strong enough to override the fact that their individual contributions—for example, by driving their car or even flying—is vanishingly small compared to the total effect. They may judge and feel the collective to be shameful and fundamentally flawed.

The assumption that humanity is fundamentally flawed and thus shameful is a trope existing well beyond the everyday environmentalist literature discussed in Chapter 2. Since claims that collective shame exist are likely more controversial than others in this book, I offer some additional signs of this trend. Consider, for example, the popular non-fiction book *The World Without Us*, by Alan Weisman, and the 2008 Discovery Channel special and the 2009–10 TV series based upon it. They paint a vivid picture of what life will be like decades, centuries, or longer after an imagined moment in which all people vanished (de Vries 2008, 2009, 2010; Weisman 2007). The popularity of this book and TV shows suggests that people are intrigued by the possibility that life will survive without us, and even that it may be better off without us. Widespread discussion of nature's revival during the stay-at-home orders in response to the COVID-19 pandemic support this claim.

Indeed, environmental literature frequently presumes that contemporary industrialized humans or, more broadly, the human species, is problematic, thus implicating a collective. While many environmentalists writing for a secular audience are hesitant to make explicit metaphysical or ontological claims about human nature, the particular words that they use often reveal their positions. For instance, describing humanity as a "parasite" (Knight), "plague" (Gray 2002, 6–8; Gray 2013; McCann 2011), "homo rapiens" (Gray 2002, 151, 184), or "cancer" (Benford 1995; Cafaro 2015; Forencich 1992; Hern 1993a, 1993b; MacDougall 1996; McCann 2011) is common in environmental literature. Authors often use these terms to denote sheer negativity, though biologists may not give the terms such a valence (Vannier-Santos and Lenzi 2011; Weinersmith and Earley 2016). Warren M. Hern carefully demonstrates how humans fit the biological definition of cancer when applied to the Earth's ecosystems (Hern 1993a, 1993b). Whether authors use such language colloquially or scientifically, they imply a judgment about the basic state of the human species—that it is human nature to proliferate unchecked, use ever more resources, and crowd out or overtly kill other species until the ecosystem is irrevocably destroyed. They imply that this is our very nature. E. O. Wilson, for instance, has written "Darwin's dice have rolled badly for Earth. It was a misfortune for the living world in particular, many scientists believe, that a carnivorous primate and not some more benign form of animal made the breakthrough" (Wilson 1993a, 1993b). Insofar as these authors assume that such destruction is wrong, our nature is too. We are in a state of shame.[1]

Phrases common in environmental literature like "human destruction of nature" imply that humans, or modern Western, or industrialized culture are separated from nature and have the ability to destroy what it should be. As environmental ethicist Lisa Gerber points out, this judgment common in wilderness literature, which has often assumed that humans are "destructive and unstoppable," that we "taint the wilderness," and that "humans cannot relate to nature in a way that is not negative" (Gerber 2002, 46, 48, 52). Simon P. James traces this vision back to John Muir, often considered one of the grandfathers of the environmental movement, who wrote about the inherent dirtiness of people contrasted with the cleanliness and order of nature (James 2014). In this train of thought, to preserve and appreciate nature is to take ourselves, or at least our material products, technology, and culture, out of nature because we are so problematic.

Recent literature about the Anthropocene often returns to and amplifies this traditional environmental theme. The Anthropocene is a geological era distinguished from others by humans acting as a geological force, changing global earth systems in such a way that will leave an impact in the fossil record (Crutzen and Stoermer 2000). The rise of agriculture, the industrial revolution, or nuclear technology are the most common contenders for the beginning of the Anthropocene, with geologists now focusing on the latter (Crutzen and Steffen 2003; Ruddiman 2013; Waters et al. 2014; Zalasiewicz et al. 2015). Value judgments about whether the Anthropocene is problematic and whether its negative effects should be countered through intentionally intervening in earth systems or

[1] A conceptual tension arises when environmentalists assume that humans are problematic to the core. If it is the case that we inevitably increase our population beyond the carrying capacities of our habitat even when we know what we are doing is harming other people and biota, one has to wonder if we can actually do anything differently. Many environmentalists gloss over this point, as they judge humanity horrible but are also incredibly optimistic about our ability to learn or change our values to catalyze action to reverse or at least slow the rate of anthropogenic environmental degradation. Without a theory of human nature that accounts for our limitations and capacities, such conceptual tensions are bound to arise. This is an obvious place for religious traditions to contribute to environmental ethics, as many, if not most, have an anthropology that deals with human limits.

Philosopher Philip Cafaro is one of the few scholars who has addressed this tension head-on. He posits that viewing the sixth great extinction (which humans are currently causing) as revelatory about the human species may disclose important dimensions of human nature not captured by other interpretive frameworks. Cafaro suggests an experimental test of human nature. If we are able to act to stop this extinction event, then we have shown that the human capacity for destruction is not inevitable (Cafaro 2015). Yet failing to stop the extinction does not necessarily mean that humanity is rotten to the core; environmental systems may have already been degraded past a point of no return before we changed our actions.

reducing consumptive lifestyles, pervade much scientific literature and humanistic studies of the Anthropocene (Northcott 2017; Rockstrom et al. 2009; Steffen et al. 2011a). To the extent that our nature is deemed problematic and could be or imagined to be otherwise, Anthropocene discourse indicates that humanity itself is shameful.[2]

In these various strands of academic literature, judgments about humanity's problematic actions (guilt) or problematic nature (shame) are more prevalent than explicit displays of collective guilt or shame emotions. That is not surprising given that academics tend to avoid overly emotional writing. Yet, exhortations to take action in these texts signal judgments of guilt (Crutzen and Steffen 2003, 254, 256; Steffen et al. 2011b). Additionally, it is difficult to imagine people avoiding feeling any guilt or shame if they presume that the human species is problematic or write about anthropogenic actions with such drastic consequences.

Clear evidence of collective emotions is found in surveys by Ferguson and Branscombe (2010) and Onwezen (2015), as discussed in Chapter 2. Recall that collective guilt and shame, among other emotions, were shown to influence the degree to which consumers were likely to purchase organic food (Onwezen 2015). People who felt collectively guilty about climate change and those who felt more guilt were more likely to claim willingness to participate in mitigation activities (Ferguson and Branscombe 2010, 138–9).

Collective environmental guilt and shame are also indicated by actions taken to facilitate restorative environmental justice between collectives. Restorative justice involves building relationships between perpetrators and victims such that they can recognize each other's value and shaping policy, practice, and habits of thought to address historic injustices and avoid repeating them. While restorative justice practices do not necessarily involve guilt and shame feelings, given the fact that these practices focus on emotions, relationships, and identity, often including apologies that encompass all of these dimensions, they are a likely place to find such evidence. In Chapter 9 I examine one such case, that of veterans apologizing to Native Americans at the Standing Rock Sioux reservation for the environmental and social harms done by the U.S. government to Indigenous Americans. Here veterans were clearly apologizing as a part of and on behalf of one or more collectives (the U.S. Army and government) to a collective, the Standing Rock Sioux. The collective guilt and shame feelings were palpable in their words, tone of voice,

[2] Many scholars rightfully argue that humans have not equally perpetuated anthropogenic environmental change. Yet if such scholars focus exclusively on nuancing claims about the causes of anthropogenic environmental change, they may overlook the fact that the Anthropocene literature's analysis of human nature is as much about our capacity as a species as it is about who has done what. Indeed, for some authors, the scale of climate change in particular provokes existential concern or shame about human nature as such. Particularly if they are widespread, these reactions should not be ignored, given the many negative effects of shame.

and bodily actions as preserved on video, in still photographs, and in interviews after the fact.

Additionally, the effectiveness of naming and shaming campaigns to change corporate and governmental behavior, as described in the next section, likely also indicate collective feelings of guilt and shame at least some of the time. While some collectives certainly only or primarily change their ways in response to such campaigns out of concern for their bottom line, collectives targeted for naming and shaming campaigns may also feel guilt and/or shame.

Finally, and probably most importantly, when we consider whether collective environmental guilt and shame exist in the contemporary United States, we should not view the examples above as isolated cases. Reading them together signals something larger—an emerging, diffuse collective that has environmental and social values, recognizes its wrongs regarding its environmental impact, judges itself to be guilty and/or shameful, and feels these moral emotions about the collective. Individuals (environmental experts cited above and everyday environmentalists) catch and get swept up in these collective emotions through interaction with each other, through texts, and through reflection on experience. The emergence of this collective emotion often co-occurs with individual guilt and shame, but is distinct from it insofar as it is felt about and by a collective. Granted, such collective moral emotions may not be felt by the entire collective causally responsible for a particular wrong. While some would say this means the emotion is not properly collective, I maintain that a smaller collective more in tune with the existence of and ramifications of a larger collective's acts may be the one to feel guilt or shame. There is a parallel here to the complicated way in which an individual may feel somewhat but not completely guilty or shameful. For instance, a person feels guilty and shameful for flying to a long-anticipated vacation and also feels justified at taking such a trip given that they reduce their travel to few such trips and that they produce many fewer emissions than their neighbors. Since individual humans are not necessarily fully consistent in their emotions, expecting a more unified experience among collectives, especially diffuse collectives, seems inconsistent at best and at worst prejudiced against collectives.

The fact that we can see evidence of the emergence of collective environmental guilt and shame at all is significant, given the barriers to recognizing such trends. Our individualistic cultural presuppositions and language make it difficult to track collectives. Because environmental values are controversial in our culture, one might not expect there to be enough environmental concern to yield collective emotions. Additionally, evidence about shame is particularly difficult to gather directly, as agents who feel it are likely to deny the feeling or problem that caused it rather than confront, admit to, or deal with the issue. Scholars who engage with shame may not name it as such (Scheff 2000). Yet, the evidence for collective environmental guilt and shame from multiple distinct kinds of sources (popular literature, academic environmental literature, social psychological research, collective actions) indicates that collective states of environmental guilt and shame and feelings on the part of multiple collectives are regularly experienced in the contemporary United States, and thus this part of the typology also captures environmental trends.

5.4 Outsiders Inducing Guilt and Shame

As much as we have focused on the internally assessed or experienced guilt or shame, we must note that other agents often catalyze such feelings and judgments. This section will explore how the induction of guilt and/or shame may happen intentionally or unintentionally among friends, family members, or coworkers; between authors and readers; and between collectives, as when nongovernmental organizations shame corporations and nations.

Family, friends, or coworkers often catalyze guilt or shame in others. Sometimes this occurs when educational attempts get internalized as guilt or shame (von Jessen 2009, 161). Moral outrage by neighbors may put pressure on others (Williams 2007). Authors of environmental self-help books or memoirs often report feeling guilty or ashamed after being judged (explicitly or implicitly) by their associates. Some of the clearest, most extended reflections on the subject come in Robyn Harding's *Mom, Will This Chicken Give Me Man Boobs?* She describes how after moving to a new, more environmentally aware city, other moms and neighbors caused her to feel bad for not buying organic food or going to an environmental rally. Harding does not think that they meant to make her feel bad, but that is the result. It is also unclear whether her neighbors meant to encourage her to take deeper environmental action, or were just commenting on her behavior. Though she did not like feeling guilt, shame, or pressure to determine how to act given complex situations and conflicting priorities, over time, Harding changed her actions and started nudging her friends and family members to do the same (Harding 2009, 46–7).

Some environmental writers aim to elicit a guilt or shame response in their readers in hopes of changing other people's minds, hearts, and behaviors. Once in a while, authors explicitly name such a motivation. For example, blogger Gwen writes in her blog purpose statement: "This blog is definitely a propaganda tool, and I am trying to make you change your ideology and behavior and essentially make you green like me (sorry)" (Gwen 2009b). Throughout her blog, she talks about her green guilt and "eco-sins" and is eager to help her readers recognize when they similarly have sinned, to have an emotional response to such transgressions, and to transform their identity into one that "sins" less and has hope of doing better. Again, this focus on identity reveals the presence of shame, even if stopping there is not the end goal of Gwen's work.

Mike Berners-Lee admits to deliberately use shocking examples to incite change about carbon emissions. Rather than general statistics about tons of emissions, degrees of temperature rise, or market prices, Berners-Lee begins his discussion of carbon offsets with the question "How many tons for a life or a death?" (Berners-Lee 2011, 13–15). He realizes that people understand "local and visible" impacts more easily than the diffuse long-term effects of their actions. "Perhaps we need to find it as shocking when we see dispersed suffering inflicted through needless carbon emissions as it would be to see the same suffering inflicted all in one place in front of our eyes by, let's say, a street stabbing." Using a "back of the envelope" calculation, he found that there is approximately "150 tones CO_2e per climate change-related death." If one follows his calculations of the emissions footprint of the average U.S. American, and takes oneself to be average, then someone dies approximately every five years from one's consumption (13–15). Then he discusses the costs of offsetting one's emissions, noting that the full value of offsets is much higher than the monetary cost. Throughout the introduction to his book, his references to death, stabbings, and other "shocking" incidents indicate that Berners-Lee wants to excite people's emotions, probably including their guilt and/or shame, while educating them.

Reporters often assume that people will have reactions of guilt or shame when reading about environmental issues. The first line of a *Business Travel* article about buying carbon offsets for plane travel reads, "Do you feel guilty about global warming every time you get behind the wheel of your car? If you are a frequent flier, start feeling more guilty" (Rijnen 2003). Here the author presumes that learning facts about (1) emissions from flights compared to other activities and (2) the ways that other travelers are offsetting such emissions will change readers' assessments of their actions or selves. The author assumes that the emotional response is inevitable, a sign of the ubiquity of environmental guilt and shame.

Indeed, environmental shaming, or at least anxiety about it, seems quite prevalent in the experience of everyday people, as indicated by the number of jokes about it (Armisen et al. 2012; Bagdonas 2014; Darkow 2016; Hamilton 2008). A significant portion of humor about environmental guilt and shame is about how people have such feelings or try to elicit them in others. In a 2016 comic, a grocery cashier reveals the store's strategy to use guilt to promote paper over plastic bag use (Darkow 2016). A similar effect is reached in a *Portlandia* sketch in which a grocery cashier and manager have an extreme reaction when a customer forgets his reusable bag, though the employees are not as self-reflective or explicit about their intentions. The employees are dumbfounded by the customer's forgetfulness. While smiling and appearing friendly, the cashier and manager do many things that would be expected to elicit shame including expressing extreme surprise that anyone could forget a bag, questioning, scolding, shaking their finger at the customer, and painting his behavior as abnormal. The cashier and then the manager liberally use the PA system, first to call for help, then as the situation escalates, to make announcements including "Man without bags here on register two." The customer offers rationalizations (he came from a friend's house and did not return home first to get a bag, etc.) and attempts to deflect attention from himself by claiming he could not be the first person in this situation. The manager replies, "It doesn't matter what you're doing. When I wake up in the morning my eyes don't forget to open, my heart doesn't forget to beat," insinuating that remembering a reusable bag is an integral part of being a human (Armisen et al. 2012). Certainly, this comment is intended to elicit shame. Throughout this sketch, exaggerated as it is, we see the customer being judged for his actions as well as for his basic being. While he does not seem to fully accept that he is as bad as the cashier and manager make him out to be, he is affected by their shaming, as indicated by his choice of words and tone, body language, and actions. Furthermore, the segment seems designed to make the audience recall similar feelings of guilt, shame, and embarrassment and also to question the extreme expectations of some environmentalists.

The prospect of being a target of attempts to induce guilt or shame is also a frequent topic of discussion in online forums. Respondents, especially in general news sites as opposed to explicitly environmental sites, often react with hostility to the authors' supposed desire to induce guilt or shame in readers, even if the authors seemed to be reporting with the most general, non-emotive language or just expressed a desire to share their own experience. For example, in response to a *USA Today* article reporting on a green guilt poll, rudyalso writes, "Green guilt? Another 'guilt' to add to the growing list of 'guilts' we the people must feel to satisfy the various extremist busybodies who feel they know better than we how to run our lives? Give me a break!" (Kornblum 2008). Rudyalso immediately assumes that because others experience such guilt that they want him or her to feel it too, though this article was extremely objective in its reporting.

Additionally, committed lay or professional environmentalists may find the stereotype damaging. James Turner, a Greenpeace employee, reflects upon his reaction to his friend's report on

> a camping trip in the Sierra Nevada. His eyes shone as he described the opalescent sky, the vitality of wildlife in spring and the fun he'd had playing with his two young daughters during the mellow evenings. It had been a really good trip, an experience to treasure, he said.
>
> I casually asked how long it took to get there. "Oh, it wasn't too bad," he said, and then caught himself, as if he'd said something wrong. "But we took the minivan this time, which I suppose means we weren't so in tune with nature after all."
>
> I felt slightly hurt. I am an environmentalist—I work for Greenpeace. Did he think that makes me some moral arbiter of fun, sternly passing judgment on

> those who ignore the perils of climate change to enjoy a weekend in the mountains? (Turner 2013)

Probably. Many people expect environmentalists to be highly judgmental, as demonstrated in the comics and comedy described above. Whether or not this hurts environmentalists themselves is a little-explored topic; more scholars have examined whether negative and positive environmental campaigns yield different results.

In all of the above cases, the primary judge and subject or recipient of the judgment is an individual, or in some cases a small group of individuals, e.g., the cashier and manager in the *Portlandia* sketch. The focus of the judgment is either the action that the particular individual did or the person's basic existence. Yet, the phenomenon of inducing guilt and shame is not limited to individual actors.

Collective shaming is quite common, as indicated by Amanda Murdie and Johannes Urpelainen's study of the shaming techniques of 2,000 international nongovernmental organizations against corporations and nations due to their environmental track records (Murdie and Urpelainen 2015). This practice, known as "naming and shaming," involves "exposing, publicizing and condemning" those who violate norms (Katzenstein 2013, 1079). While Judith van Erp maintains that naming can have a neutral intent such that it may be separable from the explicit judgment of shaming (van Erp 2014, 3210), in practice, "public awareness of a disreputable act and public shaming at some point become synonymous" (Stewart 2016, 725). Naming and shaming is most commonly done by nongovernmental organizations targeting corporations. Such campaigns may use a variety of techniques including demonstrations, letters to the editor, ads, or media kits for journalists to call into question a company's reputation and, if necessary, hurt its bottom line (i.e., through boycotts, the loss of sales due to reputational decline, or the difficulty of finding investors) (Bloomfield 2014, 266–7). Many naming and shaming campaigns aim to force companies to disclose their environmental impacts in order to inform consumers and shame the corporations not doing well. For instance, Greenpeace "rat[es] consumer electronics producers'" responsibility regarding electronic waste disposal and "sham[es] firms who fail to adequately prevent waste dumping" (van Erp 2014, 3210). Mark Stephan found that "shock and shame are key motivating factors for improved environmental performance by industry" (Stephan 2002, 190).

One of the most famous naming and shaming campaigns occurred in the 1980s, when the Earth Island Institute and schoolchildren around the country shamed tuna companies for catching and killing dolphins as a byproduct of catching tuna. In response, multiple companies changed fishing practices and labeled products "dolphin safe" (Jacquet 2015a, 3–8; Murdie and Urpelainen 2015). Fishermen are far from the only targets. Luxury jewelry retailers and banks funding environmentally and social disruptive mining operations have been targeted by the "No Dirty Gold" and "Global Finance" campaigns (Bloomfield 2014). Some naming and shaming campaigns target entire industries. For example, the "Global Finance" project maintained that "banks should take responsibility for the social and environmental impacts of their investments and develop policy safeguards to ensure they are not financing projects that are needlessly destructive" (Bloomfield 2014, 266). The #NODAPL movement shamed banks financing the Dakota Access Pipeline.

While corporations and corporate leaders are the main target of naming and shaming campaigns, nations may also be in the cross-hairs of or themselves instigate shaming campaigns. NGOs and individuals from developing nations have attempted to shame developed countries about their environmental and social records, including with respect to climate change. For instance, in a press conference at the Copenhagen Climate talks in 2009, Bolivian President Evo Morales called the United States' monetary contribution to combat climate change compared to its military spending "shameful" (Democracy Now! 2009). Norwegian NGOs have attempted to shame their own government about its climate change policies (Tjernshaugen and Lee 2007, 197–9). Australia has often used shaming techniques as a part of its multifaceted attempts to convince Japan to limit or stop hunting whales (Epstein and Barclay 2013). In Brazil, the Ministry of the Environment publishes yearly lists of regions "with high annual forest lost" to pressure inhabitants to reduce deforestation (Cisneros, Zhou, and Börner 2015). Elías Cisneros, Sophie Lian Zhou, and Jan Börner conclude that these reputational effects were significant at reducing deforestation since targeted areas had significantly less deforestation than one would expect. The state governments of Victoria and New South Wales in Australia have legal provisions enabling courts to order corporations to prevent or mitigate future harm, restore ecosystems, and publicize their infraction and the consequences the corporation faces. This last, of course, is a form of shaming (Cole 2008). Judges in the United States have given shame-based sentences to corporations (Stewart 2016, 724). For instance, a judge in Massachusetts required a local ferry company which dumped human waste into waterways take out an ad in the *Boston Herald* to publicize their crime, conviction, and fine (Smith 2010b).

In sum, each dimension of the typology of guilt and shame—whether individual, membership, or collective states or feelings of guilt or shame or externally induced guilt or shame—is found within the environmental movement. With this established, we can now ask whether guilt and shame can ethically be intentionally induced at any of these levels (Chapter 7), and how to deal with these moral emotions when they occur (Chapters 8 and 9).

6

Responding to Critics of Emotions and Collectives

Before considering the ethical questions surrounding environmental guilt and shame in Chapters 7–9, I pause here to recognize how different the methods and conclusions of my first five chapters are from most environmental, especially climate, ethicists who are wary of emotions and collectives let alone their interaction. Even when climate ethicists such as Dale Jamieson, Stephen Gardiner, J. Baird Callicott, Tracy Lynn Isaacs, and Cynthia Moe-Lobeda, get close to these topics, they ultimately turn away, leaving the evidence I discuss and ethical questions I raise inaccessible to them. In this chapter, I probe the reasons climate ethicists have not paid significant attention to the emotions, especially guilt and shame, collectives, and their interaction. In doing so, I demonstrate how attending to environmental guilt and shame ultimately challenges many elements of environmental ethics. I maintain that attending to environmental guilt and shame enables a more comprehensive environmental ethic better able to grapple with the complexities of environmental issues we face and our capacity as humans to deal with them.

6.1 Critics of Emotions and Collectives

Many modern Western ethicists implicitly or explicitly reject the significance of moral emotions because they prioritize the rational and logical. A subset of this broad group sees ethics as only concerned with judgments about the right or wrongness or the blameworthiness or praiseworthiness of particular actions, not about the circumstances under which such decisions are made. Anthony Appiah, for instance, acknowledges that some people feel moral regret, what I call "guilt feelings," and that such regrets can help one seek the good. Yet, he claims that "it is not a response which has anything to do with our judgment of what the agent should have done. It belongs, if I may put it this way, to the context of assessment and not to the context of decision" (Appiah 1991, 223). Thus, Appiah rejects feelings as external to the process of determining what someone should do and indeed to the process of determining whether they have actually done wrong. While the evidence of similar moves to avoid moral emotions is enormous, I focus here on arguments from scholars interested in responsibility, guilt, and

Environmental Guilt and Shame: Signals of Individual and Collective Responsibility and the Need for Ritual Responses. Sarah E. Fredericks, Oxford University Press (2021). © Sarah E. Fredericks. DOI: 10.1093/oso/9780198842699.003.0006

shame, especially of collectives and/or those interested in climate change and emotions or collectives. By examining positions that explicitly or implicitly draw our attention away from collectives and moral emotions, especially negative emotions, I will clarify what is lost if we do not explore these phenomena.

In *Moral Responsibility in Collective Contexts*, Isaacs makes a claim similar to Appiah's but focused on climate change. She aims to determine when people are individually or collectively, "objectively" responsible for particular actions. She sees feelings of guilt as unnecessary for acknowledging responsibility (Isaacs 2011, 90–2) because, she maintains, one can ascertain an individual's or collective's objective guilt just through an analysis of intent, agency, what the agent knew, and how the agent acted. For Isaacs, setting feelings aside is necessary so that people do not focus their attention and effort on the wrong thing, i.e., unjustified feelings rather than actual, objective individual or collective responsibility. I presume she would analogously dismiss feelings of shame, though she does not address metaphysical conditions, identity, or feelings sufficiently to engage with shame directly. Isaacs follows her own recommendation in the rest of her project.

Elizabeth Cripps' *Climate Change and the Moral Agent: Individual Duties in an Interdependent World* is broadly similar to Isaac's *Moral Responsibility in Collective Contexts* in that both examine individual and collective responsibility and duties with respect to climate change, emphasizing the need to develop collectives to respond to climate change, even as such entities do not take the place of individual action.[1] Unlike Isaacs, however, Cripps grounds her work in part in the desire to ascertain how individuals can live with themselves as people who participate in actions contributing to climate change. She recognizes that people are marred or tarnished by being in a group contributing to climate change. She also admits that people may use "psychological masking techniques" such as denial to hide such unwelcome conclusions (Cripps 2013, 192). Yet, she does not think the psychological or emotional should receive much attention: "acting collectively on climate change, rather than leaving individuals to find psychological techniques to live with it, is like curing an epidemic, rather than each finding ways of putting up with it" (194). Psychological responses, presumably including responses to guilt and shame, only enter the picture in a limited way. She extends her analogy about the epidemic: "once we reach the limits of what can be cured, then there is nothing else to do but foster such palliative tools" (194). She does not recognize the significance of guilt and shame in their own right or as barriers to individual or collective action. They are secondary issues, only worthy of attention when solutions do not exist.

At first glance, environmental ethicist Jamieson may seem to consider moral emotions, in ways that offer a corrective to Isaacs and Cripps, but on closer

[1] Cripps' collectives are more similar to my membership groups than my collectives (Cripps 2013).

examination it is clear that he defaults to rational and individualistic analysis such that he does not significantly consider emotions, let alone collective emotions.[2] Additionally, Jamieson explicitly argues that people are not feeling significantly badly about climate change, allowing him to set aside the consideration of moral emotions. Given the explicit ways in which these arguments challenge the presuppositions of this book, I will treat them at length.

In *Reason in a Dark Time*, Jamieson argues that our economic theory and political systems are insufficient to handle climate change. It "overwhelms" our capacities for thought and practical reasoning. He also claims that climate change overwhelms our affective capacities (Callicott 2014, 268; Jamieson 2014, 8, 144, 165–6, 18, 178). Jamieson seems to think that affective overwhelm is insurmountable though rational overwhelm may be addressed. He maintains that our senses and ethics evolved for responding to problems of particular spatial and temporal scales among small, tight-knit groups of people, not for the scales of climate change (Jamieson 2014, 102–3, 170, 237). Thus, he states that "climate change must be thought rather than sensed," even though he recognizes that it is difficult to change thinking and to change it in ways that influence our behavior processes (103). These positions and his consistent privileging of the rational (Callicott 2014, 207, 276–87, 292–4; Jamieson 2014, 1, 150–66, 186) likely contribute to the fact that affect, emotion, and specific emotion terms do not play a major role in his work. The index of *Reason in a Dark Time* does not contain any of these terms.

Indeed, Jamieson's most substantial engagement with emotions about climate change is when he argues that people do not feel that climate change is a moral problem and do not feel bad about it or, more precisely, do not feel significantly bad enough. For instance, he writes

> If we adopt the standard of equal per capita emissions and conjoin it with some plausible assumptions about what total emissions we should allow [i.e. under 2 °C temperature rise], then virtually all Americans are human rights violators, as are most of those in the rest of the world who live middle-class (and beyond) lifestyles. Yet [. . .] most of these people do not feel like human rights violators.
> (Jamieson 2014, 156)

Jamieson's position rests on his understanding of how people judge whether something is or is not morally suspect. They use "commonsense morality" according to Jamieson, "a dynamic system of interrelated beliefs, ideals, attitudes, emotions, dispositions, and more besides" that is shared among people and "can usefully be thought of as what we learned at our parents' knees" (Jamieson 2014, 147). He maintains that this amorphous morality is generally shared, and that

[2] Callicott has extensively chronicled the "Rational Individualism" of climate ethics, particularly Jamieson's work (Callicott 2014, 207, 276–87, 292–4).

although some discrepancies about peripheral moral issues exist, this common morality can be identified and studied.[3]

"Commonsense morality," according to Jamieson, presumes that "[a] paradigm of an act that is morally suspect is one that has the following structure: An individual acting intentionally harms another individual; both the individuals and the harm are identifiable; and the individuals and the harm are closely related in time and space" (Jamieson 2014, 148). Since climate change challenges each element of this paradigm, determining whether it is morally suspect is far from straightforward if one follows the paradigm. He writes, "If I am right about this, then it is not surprising that many people do not see climate change, at least with respect to individual responsibility, as presenting a moral problem" (150). Reasoning by analogy to a set of cases, Jamieson suggests that "[e]ven for those who do see this case [anthropogenic climate change] as presenting a moral problem, the wrongness of the acts and the culpability of the agents are greatly diminished by comparison to" the paradigm case (150, 152–60).

In the introduction to *Reason in a Dark Time*, Jamieson elaborates on this point:

> Even most of us who care deeply about climate change would have to admit, if we were honest, that we do not feel like killers when we fly or drive. We may feel embarrassed, ashamed, or hypocritical, but we do not really feel that we are worse than Mafia hit men (despite the fact that we kill gratuitously while they kill in order to make a living). (Jamieson 2014, 7)

Jamieson makes several problematic moves in this analysis. First, he draws a distinction without justification as he implies that feeling like killers is necessarily separable from feeling "embarrassed, ashamed, or hypocritical," when in fact they may be related. These feelings could stem from assessments that one's actions or being was wrong or that one is like or is a killer, regardless of whether we feel worse than a killer.

Secondly, Jamieson assumes that what he feels is what most people who care deeply about climate change feel and that what he thinks people should feel has implications for their assessments and feelings. This claim comes from his presuppositions about commonsense morality and, presumably, his assumption that emotions have a rational basis. If everyone or nearly everyone has the same morality and has a rational basis for their emotions, then it seems he reasons, that they should feel the same things about moral issues. If Jamieson does not feel a certain way, others generally would not. These are at least partly empirical claims.

[3] Nearly two decades before *Reason in a Dark Time*, Jamieson defended commonsense morality in more detail (Jamieson 1996, 324–5). However, his assumptions that morality is largely homogeneous cause him to miss the possibility (a) that more promising alternatives to the dominant ethical system may exist and (b) that the power of people who hold the type of ethics he describes to be commonsense morality exacerbates its problems.

I am not convinced, however, that the individualistic assumptions of his commonsense morality hold. Think, for instance, of Confucian, Buddhist, and some Indigenous thought, not to mention remnants of premodern Western thought that still influence people today, which challenge the individualistic assumptions of his paradigm case, as does some of the evidence of Chapters 2 and 5. His assumption that commonsense morality is shared by all prevents him from realizing these challenges to individualism. If one did not seriously question individual responsibility for climate change as Jamieson does and/or was willing to consider collective responsibility as he does not (Jamieson 2014, 1, 9, 159, 166, 176, 186, 201–37; 2015), then one may well assess one's self to be in a state of guilt or shame and feel guilty or shameful about it, as evidence in Chapters 2 and 5 indicates. Recall, for instance, Fritzie von Jessen's question "Did I just kill a penguin?" (von Jessen 2009, 16–17).

Jamieson also implies that while some people have some feelings about climate change, they are not the right or significant enough feelings. He assumes that for a feeling about climate change to be significant, one has to feel like a human rights violator, killer, or worse than a Mafia hitman. He assumes, following Daniel Gilbert, that if people cared about climate change, millions would protest in the streets, and the fact that people do not take such action is a signal that they do not care (Jamieson 2014, 168).[4] He seems to assume (a) that emotions will lead to actions and (b) that emotions and actions will be proportionate in size. Since there have not been sufficiently large climate actions to stop anthropogenic climate change, it seems (he reasons) that there are not sufficiently strong emotions about it. I think he sets the bar so high because of his own moral commitments about the seriousness of climate change and his apparent presuppositions that moral emotions need to be of proportionate strength to the danger at hand. While this may be the case ideally, assuming so relies on further presuppositions: that moral emotions and the ideals to which they are linked are always clear rather than experienced by complex, fallible people with competing emotions and reasons.

If we look at moral history, it seems that moral change is often gradual along several dimensions: in the response to new ethical challenges through the extension of existing ethical systems and/or the development of new dimensions of ethics; in the number of people who adhere to the new moral system; and in the ways these new moral considerations are ranked among others and the depth to which their attendant emotions are felt. Moral responses to sexism, racism, and

[4] The fact that the People's Climate March had over 300,000 participants in New York City (Banerjee 2014) in 2014, the year *Reason in a Dark Time* was published, and that since then multiple climate actions have involved hundreds of thousands to millions of participants indicates, by Jamieson's metric, that people do care, though action has still been insufficient (Adam and Siddiqui 2015; Bogel-Burroughs and Davenport 2019; Kaplan and Dennis 2019; Sengupta 2019; Stone 2015; Taylor, Watts, and Barlett 2019; Zhou 2018).

species-ism are just some examples of slow change. By focusing on "Why the Struggle against Climate Change Failed,"[5] as exemplified in "commonsense morality," Jamieson misses the opportunity to see emergent and existent alternatives that may address some of the moral psychological problems of climate change. Admittedly, Jamieson notes in earlier work that ethics can and do change (Jamieson 1996, 324). He implies as much when he ends *Reason in a Dark Time* with hope: specifically, "We can hope that the challenge [of climate change] will bring us together and that we will develop new conceptions of responsibility that are robust over the long term" (Jamieson 2014, 237). Yet, because he focuses on a particular diagnosis of the problem and presumes a shared morality, he is unable to see where new moral responses may already be developing.

If we are interested not just in why climate action failed but in how to conceive of, feel, and act regarding climate change, we may need to pay attention to the philosophy of people who take it morally seriously and are able to break through the impasses of Jamieson's commonsense morality. I suggest that the everyday environmentalists discussed in Chapters 2 and 5 are able to do this through their emotions and willingness to consider collective agency (albeit imperfectly and inconsistently). Jamieson, however, is unlikely to see these possibilities, given his focus on rational, individual analysis.

Even scholars open to considering moral emotions about collectives at some level have often, in the end, limited their attention to emotions to focus on rational analysis. For instance, in *The Question of German Guilt*, Karl Jaspers explicitly or implicitly acknowledges the importance of feeling guilt or of feelings in general (Jaspers 2001, 23, 66–8, 91–2, 99, 102–4, 108–9). The book is infused with a passionate concern about the experiences of Jaspers' fellow Germans, including their interaction with other nations, their judgments of themselves, their moral standing, and their emotional experiences. One of his most extended discussions of emotions occurs in the section about metaphysical guilt. He reflects on the experience of coming to recognize one's own metaphysical guilt after recognizing atrocities committed by people with whom one identifies. Jaspers writes, "By our feeling of collective guilt we feel the entire task of renewing human existence from its origin—the task which is given to all men on earth but which appears more urgently, more perceptibly, as decisively as all existence, when its own guilt brings a people face to face with nothingness" (Jaspers 2001, 75). After approaching the abyss, Jaspers retreats: "As a philosopher I now seem to have strayed completely into the realm of feeling and to have abandoned conception" (75). Then he briefly reflects on the limitations of language and analysis before beginning a new chapter in a more analytical mood.

[5] Part of the subtitle of *Reason in a Dark Time*.

Reflecting on the rise of and response to the Third Reich, he argues that prevalent emotions, especially "of pride, of despair, of indignation, of defiance, of revenge, [and] of scorn" may obscure the truth of the situation and lead to the characterizing of whole peoples as pariahs (Jaspers 2001, 6). Similar situations, he argues, catalyzed World War II and German treatment of Jewish people. Thus he is loath to overemphasize emotion (21, 34). Certainly, historical cases where unchecked emotions helped fan the flames of violence are cautionary tales against privileging emotion. Yet, one can go too far in de-emphasizing emotions. This is particularly easy given the incredible dearth of emotion in theories of ethics in the twentieth and twenty-first centuries, including in environmental ethics. To avoid emotion altogether is to ignore a part of the human experience, a situation that almost assuredly leads to an incomplete ethic. Additionally, in the case of shame it can lead to avoidance and then the exacerbation of the problem—a serious risk given the severity of anthropogenic environmental degradation and the ways in which complacency, business-as-usual policies, and denial hinder action (Gardiner 2011; Moe-Lobeda 2015; Norgaard 2011).

Jaspers also limits his study of emotions because he recognizes that language itself is insufficient for accounting for the felt experience of guilt including the "abyss" of metaphysical guilt (Jaspers 2001, 75). He seems to see his task as a philosopher to be that of articulating and analyzing concepts and philosophical arguments, even though he knows that these distinctions are not absolute (75). Given the limitations of language, Jaspers maintains that he cannot do much more than acknowledge the problem or else respond to particular instances of guilt (Jaspers 2002, 255). Certainly, language will never be sufficient for this task and yet more clarity about the interlocking pieces of metaphysical guilt (my notion of shame) is possible and helpful. Even if we are not able to fully understand the experience of shame because it is about the self, not just an action and involves experience that is not fully describable or analyzeable, we can and should analyze its ramifications, their ethical implications, and the social, personal, existential, and ethical ramifications of responding to these feelings.

For some scholars, the rejection of collective feelings of guilt or shame hinges as much on their collective nature as on the fact that they are emotions. Again, Jaspers is a prime example, as he is willing to consider both emotions and collectives but does not go as far as to acknowledge the feelings of the collective as collective. For him, many individuals indicate their feeling of the collective when they claim " 'I am Germany.' " He asserts: "The German character has no other form than these individuals. Hence the demands of transmutation, of rebirth, of rejection of evil are made of the nation in the form of demands from each individual" (Jaspers 2001, 74). This is membership not collective guilt. Admittedly, he does assert that metaphysical guilt is "not limited to the actions and lives of individual men. Rather, it refers to humanity as a whole, of which every one of us is a part" (Jaspers 1969, 52–3). This view, however, is one concerning the character or status of being human and how that shapes us as individuals, rather than guilt or status of the collective action.

Turning to environmental ethics, we see that those scholars who significantly consider emotions often do so through a particular lens: they presume that having or developing moral emotions about other entities is key to acting ethically toward them. The vast majority of such work focuses exclusively on positive emotions such as care, concern, and love.

Callicott's ethics as a whole, and particularly his major work engaging with climate change, *Thinking Like A Planet*, follows this trend. His use of Humean sentiments as a ground for his ethic and his criticism of unilaterally rational ethics make his work sympathetic to a project such as mine, which takes emotions as relevant to ethical practice and thought (Callicott 2014, 111, 292–3; 2017a, 84). Callicott, however, almost exclusively considers positive sentiments such as care and concern. He is likely motivated in part by practical concerns about how to motivate environmental behavior; he also supports his approach with sociobiological studies of altruism (Callicott 2014, 116–22). Regardless of the motivation, he does not examine negative sentiments to any significant degree, let alone the potentially beneficial aspects of negative experiences. Indeed, he seems to explicitly guard against the consideration of negative emotions when he writes "collective emotivism is not about onerous duties but such things as love and care" (Callicott 2017a, 84). He only mentions guilt briefly twice: in a list of possible emotions when discussing how sociobiology can account for altruistic behavior and feelings (Callicott 2014, 118–19) and, more importantly for our case, when he rejects "guilt-tripping" and other approaches to climate ethics in favor of moral sentiment based on the "case for responsibility to future generations" (297). Callicott does not substantially use the term "shame," though his rejection of guilt-tripping and of Gardiner's scolding of U.S. Americans as he points out the moral corruption of climate change indicates discomfort with, if not a rejection of, shaming (297).

Elsewhere, however, Callicott is interested in the moral judgments others may make about us. He writes about "the future of one's native landscape, ecosystem, and biotic community; one's inherited culture and civilization; one's nation-state and its political subdivisions after one's personal life is over," by noting: "Many people *do* care about all such things. And everyone *should* care about them lest one be judged by oneself as well as by others to be an emotionally deficient human being" (Callicott 2014, 295). These brief exchanges about ideas adjacent to shame and/or guilt are, for Callicott, just brief notes rather than fully fleshed-out elements of his system. While he advocates for a sentiment-driven ethic with resonances in virtue ethics, Callicott does not examine the negative moral emotions or how one should cope if one feels this way. Such absences should not be surprising given that Callicott focuses more on the moral patient than moral agent, on the care and concern one should have for holistic entities rather than the way emotions are linked to moral agency, and on positive rather than negative experiences.

I agree that it is problematic if environmentalists, including ethicists, emphasize blame for environmental degradation above all else. Doing so is a limited approach to environmental ethics and can be quite discouraging. A positive ethic in the sense of being both prescriptive and uplifting, or at least offering some hope, can be quite helpful in motivating behavioral change. However, examining only positive moral emotions will also lead to an impoverished ethic, as it will ignore an important part of human experience, a potential source of ethics, and/or potential hindrance to ethical life. I am interested in ethics for an imperfect world, which requires us to grapple with how to do ethics given our finitude and failings, not just our possibilities. Furthermore, when one focuses on positive emotions to have toward other entities, as Callicott does, one's analysis tends to emphasize questions of which entities are worthy of concern and why. Studying whether and how the moral emotions can inform and facilitate the class of moral patients does not mean that we should overlook the ethical questions raised by moral emotions, such as whether they are justified, whether one ought to aim to induce them in others, and how individuals and societies should respond when agents experience emotions like guilt and shame. Expanding the class of moral patients to other individual biota, to collectives like the Earth or human civilization, is an important part of environmental ethics, but emphasizing this, as Callicott does (Callicott 2014, 292–7, 302), can leave questions about whether there are also collective moral agents underexplored.[6]

When negative moral emotions are given short shrift, their likely consequences cannot be investigated and so we cannot articulate how best to respond to them. For instance, Gardiner finds it problematic that question "Who—ethically speaking—are we?" have been neglected in environmental thought (Gardiner 2012, 182). In contrast, he maintains that it is an important part of ethical analyses in general and may "facilitate much-needed progress on action" (183). He considers that notions of "recklessness," "callousness," and "shallowness," i.e. our characteristics, may shape how we conceive of environmental crises, particularly geo-engineering, the large-scale, intentional modification of earth systems to mitigate anthropogenic climate change (184–5). The most significant aspect of this argument for our discussion is his investigation of marring, blighting, and tarnishing. For Gardiner, "a life may be morally compromised, or 'tarnished,'" by one or more actions, "which [result] in a serious negative moral assessment of that agent's life considered as a whole" (189). Interestingly, Gardiner consistently uses the language of an agent's life being tarnished rather than their self or existence, I

[6] Callicott's work points toward the possibility of collectives as I conceive of them, particularly when he discusses the ontology of collectives; he does not explore such possibilities in detail (Callicott 2014; 2017b, 146).

think to keep the analysis one step removed from metaphysics and other controversial claims, focusing instead on actions (Gardiner 2011, 57, 145–8, 215, 218–19, 250, 390, 392–3). Yet, he does not recognize that participation in tarnishing or marring acts may evoke emotional responses or existential crises in a person, let alone offer ways of responding to them.[7] The ways that such marring, perceived by the agent experiencing it or others, may affect an agent's actions is only barely hinted at in his work as he acknowledges that "tarnishing, blighting, and marring" "have the potential to exert a strong pull on agents" (Gardiner 2012, 189). Similarly, as the individual agent is the basis of his analysis, including his analysis of types of tarnish, he does not consider the possibility of collective emotions, let alone their effects or the need to respond to them (Gardiner 2011, 385–7). Even when he acknowledges that individuals can be motivated by their psychological responses to what their group has done though they did not directly participate, as with South Africans or Germans after their nation's atrocities, he focuses on individuals. Here, he uses the delegated responsibility model, arguing that people feel shame because of what was done "in their name" (Gardiner 2012, 191; 2017, 31–3).[8]

Insofar as Gardiner draws an analogy between making a decision to pursue geo-engineering and Sophie's choice (of which, if either, of her children should be saved from the Nazis—a choice which results in suicide), he obliquely hints at the stakes of the experience of environmental guilt or shame about environmental action. He does not, however, address how agents might deal with such horrors, because he does not sufficiently explore the emotional or metaphysical aspects of tarnishing acts (Gardiner 2011, 385–9).

6.2 Advantages of Considering Emotions and Collectives

Some other scholars, particularly those more explicit about emotions, existential questions, and metaphysics (often scholars dealing with religion), consider the emotional and/or metaphysical aspects of tarnishing acts more substantially. Moe-Lobeda, a Christian environmental ethicist, considers emotions when writing about systematic violence and the unjust economic and environmental effects of the actions of privileged people such as herself. She writes, "To acknowledge

[7] This neglect is likely exacerbated by the fact that Gardiner focuses his attention on tarnishing with respect to geo-engineering, which is still a rather open question, rather than climate change, to which people already significantly contribute. As Gardiner's work is a first step in these directions, he cannot be expected to identify, let alone address, all of the important questions, but we can and should now go further. Particularly, we need to recognize and start addressing the question of how to deal existentially with our participation in climate change and, if we choose to undertake it, geo-engineering (Fredericks 2016).

[8] Similarly, Gardiner focuses on individuals when considering why institutions have not sufficiently responded to climate change (Gardiner 2011, 31–3, 58, 61, 215–17, 230, 403–4, 432, 439).

the widespread suffering that may be linked to my material abundance would be tormenting. How could I live with the knowledge if I truly took it in?" (Moe-Lobeda 2017a, 533). Indeed, she frequently raises such issues regarding many environmental-social issues including climate change:

> We run away from awareness of our implication in structural brutality. We run with body, heart, and clever mental manipulations. The reality would be too horrible, my guilt and grief too great. On a level too deep for words, I fear being shattered by them. Denial or its companions, defensiveness and self-justification, race in to shield us from a vision too appalling to face. (Moe-Lobeda 2013, 96)[9]

We saw a similar depth of feeling in Chapters 2 and 5. Given the potentially high stakes of considering and feeling oneself or one's collective to be responsible for climate change and other environmental degradation, more attention to the moral emotions and their interaction with identity and action is needed in order to determine their advantages, dangers, and how we may respond to them. Insofar as such emotions occur both at the individual and collective level, we also need to consider and respond at both of these levels. Because emotions can be critical to the coherence of groups and collectives, we must consider collectives and groups alongside emotions.

Emotions help to develop, cement, and sustain relationships between agents whether through love, trust, guilt, shame, or even shared anger or righteous indignation at a common target. Emotions can both foster and rise out of solidarity found within groups. They can help avoid or repair rifts between individuals and subgroups—a contrite or embarrassed apology is seen as more sincere and more likely to provoke forgiveness or acceptance than those without such displays. Emotions shared by group members also bind the group together and help a group recognize its composition as such (Barrett 1995; Fessler 2007; Gilbert 2007; Goetz and Keltner 2007; Gruenwald, Dickerson, and Kemeny 2007; Miller 1995). Recognizing that an acquaintance is also morally outraged at the latest mass killing, or that they also experience joy when the home team wins, enables people to recognize their emotional commonalities and the metaphysical and physical status and ethical commitments related to such emotions that they share. Thus, feeling something, feeling together, and/or recognizing that others share the feeling can be a strong signal of and reinforcer of solidarity.

If moral emotions are taken seriously, then they may also be seen as a signal of a different sense of responsibility, one emerging through the experience of environmental degradation in communities. For instance, Moe-Lobeda seems to use her moral emotions as a signal of and guide to her moral assessments as well as a

[9] While she uses the term "guilt," the all-encompassing nature of her assessment aligns it with my shame (Moe-Lobeda 2016a; 2016b, 36; 2017b, 210).

tool of persuasion when writing theological environmental ethics (Moe-Lobeda 2013). Indeed, moral emotions are one of the many ways people can understand their actions as praiseworthy or blameworthy by helping them identify their ideals. Such emotions are said to fine-tune one's moral compass and may even be necessary for the development of an individual's or society's moral code (Hansen Lagattuta and Thompson 2007; Hart and Matsuba 2007; Katchadourian 2010, 252–4; Zahn-Waxler and Robinson 1995). Restricting the study of responsibility to what is necessary to logically justify it through strict notions of causality and intent may force us into a realm of abstractions distant from actual human experiences, including the potential experience of collective responsibility, guilt, and shame.

The function of emotions including negative emotions is particularly significant in rapidly changing moral circumstances such as that of contemporary anthropogenic environmental degradation, since emotions, particularly the moral emotions, can also help develop moral commitments and/or help an agent recognize as yet unacknowledged moral commitments. Sometimes moral outrage, pride, guilt, or shame may occur before one has expressly articulated a moral commitment, especially in new moral contexts. In feeling the emotion, agents may express their moral commitments (Pihlström 2011, 23–4) even if they have not been explicitly identified before. Reflecting on the emotions and how they relate to previously identified ethical commitments can help one better identify one's ethical priorities. Indeed, moral emotions speak to a capacity to discern right from wrong, to empathize with others, and to have an impetus to change bad actions and maintain good ones—characteristics of an active morality. Again, these capacities function to sustain group solidarity and repair breaks in the solidarity, making the group known to its insiders and maybe to outsiders. Thus, when we take moral emotions seriously, we once again return to how individuals relate to collectives.

Finally, when we take moral emotions and collectives seriously, as I do, then we are able to ask and at least begin to answer important ethical questions: What are the ethical ramifications of environmental guilt and shame? Should we intentionally induce them? Regardless of how these emotions come to exist, how should we respond to them? These questions drive the following chapters. Insofar as Jamieson, Gardiner, Callicott, Isaacs, and other climate ethicists do not even identify these questions, they certainly cannot address them. A more capacious vision of ethics which includes emotions and collectives enables us to address a broader scope of human experience, an indication of a more robust theory.

7
Ethics of Environmental Guilt and Shame

In recent years, scholars and activists alike have raised concerns about whether the negative "finger-pointing" tendencies of environmentalists are the best rhetorical strategies for motivating environmental action (Callicott 2014, 286; Foster 2015; Nelson 2011; Nordhaus and Shellenberger 2007; van Wieren 2013, xvii). At the same time, environmental shaming seems at least as popular, if not more so, in the environmental movement. It is used by climate activist Thunberg (2019), many nongovernmental organizations, and some governments (Brandalism 2015; Cisneros, Zhou, and Börner 2015; Cole 2008; Democracy Now! 2009; Epstein and Barclay 2013; Murdie and Urpelainen 2015; Pattison 2015), and advocated for by scholars including behavioral economist Jennifer Jacquet (Jacquet 2015a) and mathematical biologists David Rand and Martin Nowak (Rand and Nowak 2009) as well as proponents of naming-and-shaming campaigns against corporations, industries, and governments. Yet, questions of whether guilt and shame should be induced for environmental ends have received little ethical attention.

Texts by everyday environmentalists do not explicate their ethics sufficiently to answer these questions, so I look to other literature from traditions in the same cultural milieu as that of everyday environmentalists to identify ethical considerations critical to answering these questions. First, I consider one of the strongest arguments against shaming: Martha Nussbaum's argument based on dignity and power, echoed in the work of the few environmental ethicists who have considered environmental shaming. Next, I turn to pro-shaming arguments based on the urgent need to act regarding climate change and its effectiveness, particularly as articulated by behavioral economist Jennifer Jacquet. Then I examine how these concerns as well as other ethical issues (including the disproportionate shaming of certain groups, especially women, and the unique challenges of shaming collectives) are manifest in the most prominent types of environmental shaming: environmental marketing and naming-and-shaming campaigns. I conclude that the risks of shaming are so great that it cannot be used lightly, even as its benefits and apparent inevitability suggests that we cannot completely rule it out.

But first, a clarification. Throughout this chapter, I will emphasize shame over guilt. While the intentional induction of both guilt and shame raise significant ethical questions about manipulation, power, and dignity, shame magnifies these issues given its focus on identity and its tendency to hinder reparative acts. Thus, I will focus most of my analysis on shame, with the understanding that concerns

Environmental Guilt and Shame: Signals of Individual and Collective Responsibility and the Need for Ritual Responses.
Sarah E. Fredericks, Oxford University Press (2021). © Sarah E. Fredericks. DOI: 10.1093/oso/9780198842699.003.0007

about shame should also make us wary of trying to induce guilt because the experiences often co-occur.

7.1 Ethical Arguments For and Against Inducing Guilt and Shame

Let us first consider one of the strongest arguments against shaming, those articulated by Martha Nussbaum, primarily in *Hiding from Humanity: Disgust, Shame, and the Law* (Nussbaum 2001, 2004, 2006, 2013, 2016). While Nussbaum briefly consider the ethics of shaming groups and considers this a more acceptable possibility than shaming individuals (Nussbaum 2004, 211–13, 241), the vast majority of her work on shame rejects shaming individuals because individual agents are those of moral concern for her. Nussbaum's notions of primitive shame and the dignity of people ground her argument against shaming. Shame in general, according to Nussbaum, "is a painful emotion responding to a sense of failure to attain some ideal state" in which one "feels inadequate, lacking some desired type of completeness or perfection" and "pertains to the whole self, rather than a specific act of the self" (184). Primitive shame is rooted in the biological experience of being human, starting in infancy. It arises because infants want to be in control and have things go their way but cannot meet their basic needs (185, 191). Relying on psychoanalytical examples, Nussbaum maintains that adults can also experience primitive shame. When they do, they "hide from humanity" insofar as they negate, neglect, or dismiss that being human means that one is "finite" but has expectations that surpass one's finitude (173).[1] They reject the idea that individuals are limited and are dependent on others. For Nussbaum, this idealism is narcissism which should be overcome as one develops "relationship[s] of genuine mutuality" as an adult (187). Fulfilled adults, argues Nussbaum, should be beyond primitive shame because it is rooted in unrealistic expectations. The finitude of humanity does not carry moral weight for her except when people react to such finitude in inappropriate ways, i.e., by being ashamed.[2]

[1] I am wary of Nussbaum's reliance on psychoanalytic cases over social psychology because of the many critiques of psychoanalysis including the difficulty of replicating their claims using the scientific method and the degree to which conclusions about all people can hinge on the analyses of a few people. Moreover, the focus of such analysis on individuals to the exclusion of collectives is also out of step with my project. Nevertheless, her treatment of shame does helpfully explore its connection to judgments of the abnormal and the danger of shaming to human dignity.

[2] I disagree with Nussbaum's claim that overcoming primitive shame is required to be a fulfilled adult because this expectation is in and of itself a form of idealism that belies the limitations of human nature. I reserve the possibility that humans are in a state of shamefulness, and that it is reasonable to feel ashamed about this because we can imagine beyond our capacity to act and often do have some (albeit imperfect) abilities to act better than we do. I reject her tendency to categorize adults who experience primitive shame as infantile because I am not convinced that such an experience can be helped and maybe should not be. I also find her tone bordering on scorn for those who experience

Like primitive shame in infants, the societal shaming of individuals and groups is a ubiquitous part of human life arising from how people experience themselves and interact with others (Nussbaum 2004, 173–4). Individuals want to feel powerful and that they are without limits. Nussbaum argues that to achieve this, people set up "in" and "out" groups where they are "normal" and others "abnormal." Such classification often hinges on features of existence that one cannot help, such as gender, sexual orientation, or ability. Thus, shaming has roots in, exacerbates, and is exacerbated by primitive shame both of the shamer (who aims to avoid his or her limits by focusing on those of others) and of the shamed (who perceives judgments of abnormality as exacerbating his or her limits). While certain individuals are more often targets for shame, it is a possibility for everyone because "we all have weaknesses that, if known, would mark us off as in some ways 'abnormal' " (173).

For Nussbaum, being shamed and feeling shame are problematic for many reasons. Shamers often fail to stop with shame, moving from shaming to actions that belie and restrict the dignity and worth of the shamed. Shaming can give a mob power to suppress who and what they do not like. Shaming penalties are often unreliable, since they frequently are not proportionate to the problem they claim to address and/or do not "target" the right people. She also maintains that shame tends to humiliate people and put more people under social control even though it is not reliably associated with retribution or reform. Furthermore, she finds that state-sponsored shaming undercuts the ideals of liberal democratic America (Nussbaum 2004, 233–8) and that paralysis arising from shame is a problem. Her main concern is, however, that shaming is dehumanizing and against the dignity of the person (230–8). According to Nussbaum, when one is shamed, one's whole self is judged on characteristics out of one's control, whereas judgments of moral culpability and character must be based on acts or omissions within the agent's purview. Insofar as it is conceived as encompassing the whole of what a person is, shame entails a refusal to acknowledge the basic dignity and worth of the individuals being shamed. These results are problematic in and of themselves and because they devastate the individual's ability to see themselves as worthy, setting off a lifetime of "shrinking" and despair (234–7).

Intensifying these concerns is the idea that, once shamed, there may well be no going back. While Nussbaum suggests that changing beliefs can change emotions in her general discussion of emotions (Nussbaum 2004, 25), when she turns to shame this possibility receives little if any attention. It seems that she holds little hope that one's beliefs about a shameful self can change. The whole-self-judging nature of shame is so encompassing it can break one's spirit. The frequent permanence of the degradation (in the eyes of the shamer and shamee), as evidenced in

primitive shame as adults uncharitable and inconsistent, considering her significant concern for the dignity of individuals throughout *Hiding from Humanity.*

her examples of historical and contemporary shaming, provide further support to her position (230).[3]

It is true that some people may not recover from experiencing shame. Yet, Nussbaum's focus on this possibility, though well within the tradition of philosophers who examine extreme cases to delineate the boundaries of a problem, problematically constrains her study of shame (Nussbaum 2004, 193). Namely, overemphasizing the extreme cases may cause her to miss at least three important nuances of shame experiences. First, while shame and shaming often occurs about things that are out of one's control, such as gender, race, ethnicity, and height, I focus in this book on shame that occurs because of how our actions, which are at least partly under our control, reflect our identity. I consider (as in Chapter 4) a metaphysical account of humanity that allows for the possibility that some shame is a proper response to being human and explore shame arising from reflection on action.

Second, moral and social psychological accounts of guilt and shame, including extreme collective cases of shame (e.g., post World War II German experiences) and mundane (e.g., lying about spilling wine on a rug), highlight the fact that guilt and shame are often entangled with each other and with other emotions (Ferguson et al. 2007, 333). Taking this work seriously allows us to recognize that shame may involve a judgment about a person's character or identity, but still does not fully describe who the person is or who they perceive themself to be. The moral and social psychological literature allows for this possibility, as it repeatedly names shame as whole-self-judging, but does not necessarily claim this shame as the exclusive characteristic of the self. If shame is not the only characteristic of a person, they may not be completely devastated to experience it. We see such nuanced responses among everyday environmentalists whose actions prompt them to recognize their flaws of identity and that they have done and can do otherwise. Of course, the extreme form of shame is a significant risk (to which we return in the next section), but it is not the only ramification of shame. A fuller picture of shame's possibilities is needed to fully evaluate its use.

Third, Nussbaum's focus on devastating shame, particularly by the state, directs our attention away from important considerations regarding the practice of shaming in communities. She does not examine cases in which shame is averted,

[3] Jon Ronson's *So You've Been Publicly Shamed* provides additional support for Nussbaum's concerns about shaming as he chronicles the devastation that shaming in the internet age can bring. When a crowd of often anonymous people shames an individual for perceived wrongdoing, they can call such significant and rapid attention to the shamed person that they radically alter the shamed's place in society and, sometimes, self-concept. Ronson's research highlights how shaming in general and internet shaming in particular may be grossly disproportional to wrongdoing, may not be based on actual wrongdoing, and often provides no opportunity for reintegration into the community. Internet shamers often have no previous or ongoing relationship with the shamed and are not held accountable for even extreme shaming. Indeed, they frequently do not know of the devastation to which they contribute (Ronson 2015).

mitigated, or eased with time through internal resources of the shamed or social support of the shamed, as in many of the examples of environmental shame discussed in previous chapters. Consequently, she does not sufficiently consider the possibility that communities, including religious communities, can not only be a source of shaming but can also provide ways to cope with shame (whether or not it is intentionally induced), particularly when shame is accompanied by moral support, ritual action, and the development and implementation of a metaphysical anthropology to counter totalizing shame.[4] Thus, it is easy for her to view shame as primarily an all-encompassing, insurmountable experience, and to all but argue that it should not be used (Nussbaum 2004, 15).

Certainly, Nussbaum's warnings about shaming are valuable for those of us concerned with human dignity and/or liberal democratic ideals. Yet, her nearly exclusive focus on the shaming of individuals by the state and relative lack of attention to the existential and collective dimensions of shame also prompts us to look beyond her work to scholarship engaging the biological and existential dimensions of environmental shame in particular.

For instance, William R. Jordan III and co-authors denounce shaming as a tactic. They maintain that existential shame—that which is based on the fact that we must kill to eat and therefore to survive—should not be used "to manipulate others, or as a motive for behavior," and object to "the manipulative use of shame to control behavior or even to inculcate an ethic" (Jordan, Barrett, and Curtis 2012, 302). Like Nussbaum, they are concerned with the dignity of all people as well as their knowledge, agency, and ability to meaningfully participate in decision-making that affects oneself rather than falling prey to the direct or indirect power of others. The only time they are open to using shame as behavioral motivator is "to the extent that it arises from apprehension of our moral failings and limitations and so accompanies guilt" (302). Jordan and colleagues are right to be wary about the dangers of coercion or manipulation when using existential shame as a motivator. Unfortunately for our purposes, they do not discuss how to draw the line between problematic manipulation and justifiable influence, or how to deal with shame arising from guilt, because they focus on the existential and, even more so, on ecological restoration rituals particularly related to human relationships with land. I return to the issues of manipulation and power in the next two sections and to conditions that can prevent or mitigate the possibilities of manipulation in the following chapter.

Concern about shaming or inducing guilt takes a more relational and pragmatic turn in the work of Sally Bingham, an Episcopal priest and leading environmental educator and advocate. Bingham considers using guilt and shame during

[4] Nussbaum does allow for a limited positive role for shaming, as will be discussed in the next section and chapter, but does not sufficiently develop why this kind of shaming is positive and potentially justifiable, nor does she adequately explore whether it is permissible when perpetrated by nonstate collectives (Nussbaum 2004, 244–5).

sermons, as she reflects about how religious leaders can and should be "an important moral voice in the community" even as they "risk [...] alienating their parishioners if they are seen as unprofessional or as using the pulpit as a soapbox" (Bingham 2007, 153). She describes an experience in her early work on climate change in which she preached about the selfishness and sin of people who drive big cars (156). One member of the congregation walked out during the sermon. Bingham admits that she "felt badly that he was offended," but also recognizes what she seems to see as a positive result: months later the man reported that the sermon prompted him to buy a smaller car for everyday use, reserving the SUV for mountain trips. This experience taught Bingham several lessons: "First, no one likes to feel guilty (or worse, to be publicly shamed) [...] and, second, over time, the voice of the clergy does have a powerful impact" (156). While Bingham acknowledges the impact of her sermon, she does not advocate that others take her experience as a direct guide. Rather, she argues that clergy need to "respect [...] our parishioners" not "offend" them "and never be so bold and controversial that a parishioner feels the need to walk out" (156). She makes this argument because "we may not get the chance for reconciliation" (156). Bingham recognizes the possibility of narrowly defined success from inducing guilt or shame, but when push comes to shove, relationships and the dignity of people should come first. Later, she maintains that using guilt as a tool to motivate action about climate change "rarely works—not even in faith communities where congregations are used to lessons based on guilt" because "climate change is still too new and too abstract and untouchable to lay blame on individuals or communities" (161). She is also concerned that making people feel bad often turns them off such that they dismiss or ignore rather than engage with the issue. Rather than trying to induce guilt, she recommends connecting environmental problems and solutions to things people care about, whether saving money on electric bills or reducing asthma rates. Yet, she does not recognize that logically connecting another's actions and existence to environmental problems can also expose responsibility for the problems of the past and present which may in turn bring about guilt and shame even if eliciting these emotions is not the goal. Though her analysis could have gone further, Bingham's reflections reveal (1) the importance of attending to moral emotions whether or not they are intentionally elicited and (2) that the temptation to use guilt or shame can be significant even among people who are ultimately against their use.

Not all scholars and activists are, however, against shaming to motivate behavioral change. Indeed, some, particularly Jennifer Jacquet, advocate their use given the urgency of environmental problems including climate change and the concomitant need to quickly change behavior.[5] Evaluating her analysis allows us to

[5] A number other texts advocate renewed attention to shaming in order to redevelop a deeper shared sense of values in communities, to standardize values across all humans, or to curb egregious behavior. See summary by Nussbaum in (Nussbaum 2004, 227–50) as well as (Cisneros et al. 2015;

(1) chart a path between it and the absolute or near-absolute objections to inducing guilt and shame as discussed above and (2) articulate how to respond to the guilt and shame that do and will exist whether or not they are intentionally induced.

While Jacquet recognizes that shame can be a feeling, she focuses on the act of shaming or being shamed. For her, this entails exposure in front of a real or imagined audience where the individual (or group) is held to a standard in order to promote social control (Jacquet 2015a, 10–11). In contrast, her notion of guilt involves internal judgments of oneself without the need for an audience or group standard. She maintains that she does not want to investigate "how to make shaming morally correct" but rather "what makes shame effective at establishing and enforcing norms, as well as some characteristics that might make shaming more acceptable to the crowd" (99–100).

Despite her attempt to move away from ethics, Jacquet does take an ethical stance based on utility and expediency insofar as she privileges them above other sorts of considerations. She argues that new forms of shaming may be a valuable, if not necessary, method of responding to systemic problems. Jacquet is particularly interested in shaming as a solution to the challenges of the Anthropocene, in which humans are a global force and, even more importantly in her analysis, are bound together in our experience of the challenges we face, whether infectious disease or climate change, due to the global scale of the problems and the fact that perpetrators increasingly will not emerge unscathed (Jacquet 2015a, 25). Given the urgency, seriousness, and scope of climate change, Jacquet finds that changing behaviors is necessary. Yet, she does not think sufficient change will happen quickly enough if we wait for everyone, especially the most polluting, to develop and implement new or existing values.

She finds that shaming or threats of shaming can be effective and far economically and socially cheaper than other forms of punishment including physical punishment. She also notes that shaming is often accepted and that shaming groups is "probably more acceptable" than shaming individuals, because she does not think groups have the same claims to dignity as individuals (Jacquet 2015a, 13, 24, 97, 172–83). Therefore, she often prefers shaming to other means of social control and, indeed, advocates for it.

She turns to shame rather than guilt to facilitate the influence of relatively new norms because in her mind guilt depends on a person already assenting to a particular ideal and judging themselves according to it (Jacquet 2015a, 12, 85). If they have not yet adopted the new norm, then they cannot apply it internally.

Cole 2008). This literature does not consider environmental shaming, tends to focus on state shaming, and/or is looking backward to supposedly lost values. As these contexts are significantly different from that of environmental shame, which is not primarily generated by the state and is typically is thought to require new or at least an expansion of existing values, I will not examine this literature in detail.

External sanctions are needed. Shame fits the bill because it can aid the implementation of norms that have not yet been widely adopted by society. Since "people tend to calibrate their actions to what they see or hear as common behavior," when society is portrayed as supporting a norm, individuals are likely to follow the norm (78). For instance, she cites studies about how people change habits of using latrines, washing one's own dishes at the office, or recycling based on what others do (82–3). Shaming makes the new norm seem common enough that others (individuals, corporations, industries, and governments) are subsequently more likely to follow it (86).

Jacquet's detailed study of the conditions that make shame effective are one of the major strengths of her work. Shame is effective, by her account, "when people reform their behavior in response to its threat and remain part of the group" (Jacquet 2015a, 99). In other words, she articulates an action-oriented shame that can avoid paralysis (though she does not discuss this much, if at all) and guards against the shaming about what cannot be helped and ostracism that arises with shaming (and which concern Nussbaum). Jacquet elaborates by identifying seven conditions of effective shaming.[6] First, the transgression is a concern of the audience, in other words, the "audience is also the victim" (101). When this occurs, the audience is likely to care more, so that the shaming will be more effective (100–2). For instance, she argues that smokers harming bystanders are a better target of shame than an oversleeping friend.

Second, there needs to be a "big gap between desire and actual behavior" so that the shaming could significantly change behavior (Jacquet 2015a, 103). Otherwise, shaming would not be worth it. Third, effective shaming can occur when formal punishments for the infraction are not established, or when the official penalty or its enforcement is weak and the effects of shaming relatively strong. For instance, some environmental fines are so small as to insufficiently affect a polluting company. In contrast, the reputational and economic harm of calling out their infractions could be much more significant (106–7). If formal processes were established and used effectively, shaming would be unnecessary.

Fourth, the "transgressor should be sensitive to the source of the shaming" (Jacquet 2015a, 107), a condition closely related to the fifth, that the audience should "trust the source of the shaming" (111). Woven into her discussion of guideline 4 is a condition that deserves a number of its own, that since "to keep shaming sustainable into the future, it is also important that individuals or corporations who are exposed are given a chance to reintegrate with the group" (110). Pairing shame campaigns with opportunities for praise or honor may not only work better (Jacquet's main concern) but may also avoid the denigration of

[6] Jacquet actually uses the term "habits" rather than the more accurate characteristics of, conditions for, or guidelines for shaming (Jacquet 2015a, 100). E.g., her first "habit" is that the "audience is also the victim" (101). *Paying attention* to this condition could be a habit, but it itself is not a behavior.

participants (22, 110, 164–5) that she mentions and is of prime concern to Nussbaum, Jordan et al., and Bingham.

Jacquet's sixth characteristic of effective shaming is that it is "directed [to] where possible benefits are greatest" (Jacquet 2015a, 113). For instance, targeting the "ninety corporations [. . .] responsible for nearly two-thirds of the historic carbon dioxide and methane emissions" and/or "the wealthiest 10 percent of humanity [which] contributes an estimated 50 percent of global fossil fuel carbon dioxide emissions" may be a quicker way to significant progress on climate change than targeting everyone equally (66).

Finally, Jacquet maintains that effective shaming needs to be implemented "scrupulous[ly]" (Jacquet 2015a, 114–15). By this, she means it needs to occupy a space between extreme actions, neither of which is as effective as the middle ground. For instance, the "frequency of exposure" and "style" of shaming need to be those that get "attention" and are "the most acceptable" to the population (115–17). Unfortunately, she does not explore what acceptability means or whether being acceptable to the population always means that an action is praiseworthy.

Jacquet is particularly interested in shaming agents for "the transgressions that matter most, the most responsible transgressors, and perhaps which transgressors are most likely to change and which solutions are most likely to be adopted" (Jacquet 2015a, 113). Relying on behavioral economics research, she delineates guidelines for identifying an agent to shame. The best targets occur when a concentrated and narrowly defined group contributes to a problem. For instance, the supply of shark fins come from all over the world, whereas the demand is concentrated among wealthy Chinese people (113). Alternatively, most of the problems she is concerned about including climate change have supplies that are concentrated (i.e., carbon dioxide emissions in a few industries) whereas the demand is diffuse. In these cases, targeting the outsized contributors to the problem, whether individuals, corporations, or nations, can be an effective and efficient use of resources to address the issue.

Jacquet's attention to effectiveness in general and its particular characteristics are helpful for a number of reasons. First, a concern for effectiveness can counteract the overemphasis on emotion and short-term individualistic solutions that pervade many discussions of environmental challenges. In *Bag Green Guilt*, for instance, Jen Pleasants frequently emphasizes taking action in ways that may make people feel better but do comparatively little to help them be a materially effective environmentalist (Pleasants 2009). The many discussions about remembering cloth bags when one goes shopping are only sometimes accompanied by reflections on the fact that one does not need so many consumer products or that one could take political action to change infrastructure or energy use—actions with a much larger environmental impact than using a reusable bag. Concrete actions that are effective, whether working to enact laws to reduce greenhouse gases or reducing one's air travel, are critical to addressing shame because they can change how one exists and conceives of oneself, kicking one out of a shame spiral that ineffective or unrealistic actions may only exacerbate.[7] Of course, material or economic effectiveness are not the only important considerations; emotions and identity matter as well. The trick is to figure out ways of making the hidden impacts of actions visible, valued, and meaningful such that changing them has at least a similar if not greater emotional impact than the already visible everyday actions like choice of bag at the grocery store. Then the action can be effective not only emotionally and existentially but also physically.

Jacquet's consideration of effectiveness positions her as a helpful corrective to Nussbaum's considerable caution against shaming, without going so far as to advocate shaming unilaterally. While Jacquet considers shame a useful and maybe necessary strategy for changing actions, she does not advocate for it blindly, unequivocally, or in all cases. Her arguments draw on many case studies and accounts of empirical research that widen our consideration of the possibilities of shaming beyond the most dire that Nussbaum and others emphasize and beyond Nussbaum's focus on the state as primary shamer. Particularly, Jacquet broadens the conversation about shame as she recognizes that the relatively less powerful can effectively shame the powerful including nations and corporations—a possibility enhanced in this age of cheapness, accessibility, and effective internet shaming (Jacquet 2015a, 123–4; Ronson 2015). She maintains that shaming groups "probably more acceptable" than shaming individuals because they do not have dignity of their own (Jacquet 2015a, 24, 174). Her recognition of individual dignity implies that she too would have qualms about extreme shaming. She acknowledges that online shaming can pose particular risks in that with it "[t]he line between punitive and activist shaming by the crowd can be difficult to draw" and that it is easy to disproportionately punish relatively minor infractions (127). She does not, however, analyze the extra dangers of internet shaming for vulnerable groups in detail.

Jacquet emphasizes shaming groups, in part, because she recognizes that individual acts, including but not limited to consumer acts, will not be enough to respond to collective action problems, including those of the environment. Consumeristic solutions to relieve guilt, she argues, have too often shifted our focus to individual actions rather than changing laws, policy, or collective action. In her turn to groups, however, Jacquet does not go far enough. She does not consider collectives, as her work seems to stop at the level of membership. She also fails to sufficiently consider who gets to decide which norms are being inculcated through shaming, and the possibility that norms advanced through

[7] Effectiveness is also critical in responding to guilt. Guilty people motivated to take reparative action may need guidance to determine what might actually be effective according to different standards (physical, economic, political, relational, emotional, existential) across various timescales.

shaming may not address collective action problems such as climate change. Her lack of explicit engagement with ethics contributes to these limitations. Yet, a project focused on effectiveness cannot set aside ethics, as standards of effectiveness depend in part on normative values (Fredericks 2014a).

Another major limitation of Jacquet's work is that since she thinks effective shaming can occur without emotions, she presumes she is able to sidestep questions of the negative emotional experience of shame and their ramifications. She recognizes that shame feelings can occur and that it is problematic when dignity is eroded through shame (Jacquet 2015a, 10, 19), but does not sufficiently consider either that the act of shaming for social control can lead to feeling shame or the negative consequences that can stem from such feelings. When she acknowledges the pain of shame she argues that "ideally, shaming creates some friction but ultimately heals without a scar" (99), but does not fully examine how scars can be avoided or diminished if the cut is inflicted. For all of these reasons, her model needs to be supplemented by a recognition of, and ways to deal with, the potential negative effects of shaming.

Anyone considering shaming as a tool also needs to be more aware of the epistemic limits surrounding this practice. Advocating for effectiveness should not lead one to presume that one can have complete, perfect knowledge about what the results of an act of shaming will be. Certainly, some economic and psychological studies examine the effects of these characteristics of shaming, but it is far from clear that the available research is sufficient to accurately predict which conditions will be most effective in any particular case. Furthermore, such research may never be sufficient given the variety of factors that influence it: the psychological character of the people involved and relationships between them as well as existing norms, cultural practices, and patterns of assent to them. The conditions that make shaming more or less effective should be taken seriously, but making a decision to induce shame or guilt will always rely on imperfect information. This uncertainty should be recognized if one is considering inducing guilt or shame.

Finally, while I agree with Jacquet that concern for efficient and effective shaming should be a critical component of determining whether and how to induce or respond to shame (or guilt), I am concerned that Jacquet's argument overemphasizes effectiveness, and a narrow vision of effectiveness at that. Effectiveness certainly deserves more attention than environmental ethicists have traditionally given it, but we must also take care not to overemphasize efficiency to the point that (1) the value judgments efficiency depends on are obscured; (2) physical or economic effectiveness overshadows emotional, relational, or existential effectiveness; or (3) concerns about the kind of agent we become through acting should be dismissed due to prioritizing effectiveness.

Through this examination of arguments against (Nussbaum, Jordan, Bingham) and for (Jacquet) shaming, the general contours of the debate have taken shape: a concern for the dignity of individual in tension of with the prioritization of timely, effective action in the face of civilizational threats such as climate change. Many other ethical considerations about environmental shaming will come into view as we consider the most prominent types of intentional environmental shaming: environmental marketing and naming-and-shaming campaigns.

7.2 Power and Misdirection in Environmental Marketing

One of the most common and influential uses of environmental guilt and shame to intentionally provoke behavior occurs in marketing campaigns that suggest that buying a product or service can alleviate past or prevent future guilt or shame. While marketing is not intrinsically ethically problematic, environmental marketing using guilt and shame often relies on unjust power dynamics and misinformation while overemphasizing individual household consumer solutions over social or political change. Thus, while environmental marketing can catalyze environmental behavior, it raises ethical questions about power, access to information, and whose values are prioritized, as well as how to guide actions that yield effective environmental change.

While precursors to the environmental movement, particularly the hygiene and workers' rights movements, have existed since before the turn of the twentieth century (Gottleib 1993), green marketing emerged more significantly in the 1970s (Gibson 2009, 428) and blossomed in the late 1980s (Coulter and Pinto 1995, 697) and early 1990s (Tucker et al. 2012, 10; Yang, Jiménez, and Yowei 2015). Occurring through a variety of media including newspapers, magazines, billboards, TV, radio, and the internet, environmental marketing is used by corporations selling a product or service or by nongovernmental or governmental organizations trying to change perceptions and behavior. My analysis examines corporate marketing campaigns, given their prevalence, and clear illustrations of ethical challenges in green marketing using guilt and shame.

Environmental marketing campaigns use a variety of strategies, including increasing knowledge about environmental degradation or how a product, service, or activity supposedly limits such degradation; associating a product with prestige; or using emotional appeals to elicit the viewer's pride, joy, anger, fear, guilt, or shame and shape their subsequent action. Emotions are often used because some studies find that they work better than appeals to "pure facts" (Hartmann and Apaolaza-Ibanez 2012) or attempts to change environmental attitudes as a catalyst to changing action (Fraj and Martinez 2007, 31). While the literature is inconclusive about whether positive or negative emotional appeals are more effective (Onwezen 2015; Onwezen, Antonides, and Bartels 2013; Rees, Klug, and Bamberg 2015; Tilikidou and Delistavrou 2008), enough scholars and market researchers find negative emotions equally or more effective than positive emotions that the practice is common (Koenig-Lewis et al. 2014; Moon et al. 2016).

Marketing research about negative emotions uses the term "guilt" more often than "shame," often with the same connotation used in this book (action-oriented guilt versus identity-oriented shame), but many studies use only one term to encompass both meanings, or use both terms without crisply delineating their definitions (Burnett and Lunsford 1994; Laseu 1991; Qi and Roe 2016). The phenomenological entanglement of guilt and shame adds to the complexity of such work. Research generally agrees, however, that these moral emotions can shape attitudes toward products and brands as well as motivate purchasing behavior.

Using environmental guilt and shame in environmental marketing can have clear benefits. For instance, attempts to elicit guilt and/or shame may involve educating people about the environment, their interaction with it, and the self-imposed dangers of certain actions (Greene et al. 2015). Beyond the intrinsic benefit of knowledge, such campaigns may be instrumentally beneficial, as they can set the stage for environmentally friendly actions (if the agent has relevant values and the will and capacity to act) (Barbarossa and Pelsmacker 2016). For instance, environmental marketing has been shown to help reveal and prevent the root causes of food waste, saving individual consumers money, and conserve the resources used to grow, transport, and cook food (Qi and Roe 2016). Similarly, Marlene C. Onwezen demonstrates that guilt can motivate people to increase their consumption of organic food (Onwezen 2015)—a behavioral switch that can limit the exposure to pesticides of agricultural workers, consumers, and other biota while increasing the demand for organics and eventually increasing their availability and affordability.

Ideally, these cases can be ethically positive for a host of reasons: they may provide meaningful, relevant information about the environment and potential actors' behavior based on best available knowledge; they have a clear, beneficial, and effective impact on the environment; and they utilize values that appear to align with the values of the actor(s). Furthermore, the guilt and/or shame experienced may be unpleasant, but it is presumably small compared to the greater good and probably not as debilitating to the actor because of the scale of the issue, the ready solution, the ease of acting, and the support for the action. These assumptions, of course, do not always hold.

Advertisements using strong guilt and/or shame appeals have been shown to have effects opposite to what was intended. Aurélien Graton et al. note that environmental guilt's reparative function may be triggered if reparative suggestions are given in the ad (Graton, Ric, and Gonzalez 2016). Yet, when reparative suggestions were too strong or forceful, participants tended to decrease reparative actions. Viewers of ads inducing high amounts of guilt generally view the ads as more "annoying [...] anger provoking and [...] upsetting" to the point that they were not swayed by them (Jimenez and Yang 2008, 240). Intense appeals to guilt in ads may be seen as attacks on self or behavior, and may subsequently lead to a negative assessment of the ad, brand, and/or product to the point that the consumer avoids them (Jimenez and Yang 2008, 234). Similarly, other scholars have found that "shame, stigma, and guilt are powerful motivators unless they are used too much" because excessive shaming can "force [consumers] back into denial mode" (Bennett and Williams 2011, 82–3; Chan and Lau 2002; Jimenez and Yang 2008; Yang et al. 2015). Corroborating these studies is work about ads in general which has found that when ads draw attention to something negative about consumers' identity (as shame ads would do), consumers are likely to reject the ad (Thomas, Trump, and Price 2015). These results are consistent with the earlier analysis of shame, in which guilt about significant harms often bleeds into shame, which is likely to evoke hiding, denial, and a refusal to engage with the issue.

Even if the intensity of guilt or shame in any one ad is unlikely to cause a negative reaction, the cumulative effect of many ads may have a negative effect. These conditions may make people aware of an ever-increasing number of products, services, and experiences about which one may feel guilty or shame (e.g. cleaning supplies, food, energy, transportation, clothing, and indeed any consumer products). With such a long list of problems, it is easy for a person to become overwhelmed and even paralyzed by the sheer number of bad actions to rectify (guilt), by what so many infractions say about their identity (shame), or both.

Such feelings are easily exacerbated by the fact that ascertaining which goods, services, and activities are environmentally beneficial is difficult, since environmental systems are complex; multiple values may be involved; and information is hard to come by, confusing, and difficult to evaluate for reliability. All of these challenges are exacerbated for everyday environmentalists who have limited time. In the absence of standard metrics, independent monitoring, and widespread educational campaigns—all of which are expensive and require many, often hidden, value assumptions—consumers typically obtain much of their information about goods and services from marketing campaigns. The impulse of guilty agents to take reparative action may make them more vulnerable to marketing suggestions about how to act, particularly when a right course of action is unclear.

Marketing that uses emotional appeals may also backfire. Paying for environmental offsets for greenhouse gas-emitting activities such as flying is often seen as a way to assuage guilt and avoid shame. Yet, it may make people feel as if they have a license to emit rather than a responsibility to curb their behavior. Indeed, this is a common fear of critics of carbon offsets (Chea 2006; Revkin 2007; Smith 2007; Verdier 2007). While empirical studies of this potential problem are still rare, Matthew Harding and David Rapson found that households participating in a carbon offset program increased their home electricity use by 1–3 percent (Harding and Rapson 2019), while Jingru Liu and colleagues found that increased energy efficiency in home air conditioners correlated with increased energy use in China (Liu et al. 2016). Recent research about state or national schemes shows mixed results (Anderson, Field, and Mach 2017; Wang and Wang 2014). Other research about unintended consequences of policy-making regarding reputation and self-perception indicates that rebound effects should be seriously considered. For instance, a famous study imposing a fine on parents if they were late picking up their kids from daycare found that the fine led to more late parents, as the convenience of a flexible pickup time was worth more to parents than the fine. Parents assuaged their guilt by paying the fine (Bennett and Williams 2011, 82–3; Gneezy and Rustichini 2000). People who purchased green products were subsequently less likely to be generous and more likely to lie in order to earn more money. Nina Mazar and Chen-Bo Zhong concluded this was because the green behavior "establishes moral credentials that can subsequently license deviating behavior" (Mazar and Zhong 2010, 497). Given the possibility of these rebound effects, green marketing campaigns are far from straightforward solutions to environmental degradation.

Green marketing may also be misaligned with environmental aims, as it is often aimed at increasing the consumption of goods and services that are not necessarily environmentally beneficial. For example, a throwaway but "green" lifestyle is often promoted by home design, renovation, and real estate media. Such campaigns entice homeowners and property flippers to rip out functional kitchens and bathrooms because they look "dated," sending tons of material to landfills to be replaced with the latest ecofriendly bamboo flooring or recycled glass countertops. Using more durable goods, investing in timeless fashions (or not minding an older look), and repairing objects when broken or worn would often have a much smaller environmental impact in the long run, but such actions are rarely the subject of marketing campaigns, as they make less money for businesses.

The green consumerist mindset is not limited to explicit advertising. It often dominates media coverage of environmentalism in general. For example, one study of environmental articles in four women's magazines published in April, Earth month (*Glamour*, *Marie Claire*, *Self*, and *Vanity Fair*), found that only two articles about the environment "had no green commodity component at all" (Smith 2010a, 76). Given the power and ubiquity of such messaging coupled with explicit marketing, counter-narratives to consumerism have difficulty competing with it. Even if one has awareness, values, knowledge, will, and power, one still faces significant challenges to living a lifestyle not dominated by consumerism. People may not have the vocabulary to assess whether resistance may be valuable, let alone the awareness that it is possible. Problematic manipulation easily occurs when the interests of one, more powerful side determines the actions being promoted. In these cases, the values and indeed dignity of others are easy to overlook.

For instance, green marketing utilizing environmental guilt or shame often suggests that the solution to environmental problems is a product or service, not an affective, relational response. Yet as Chapters 2 and 5 suggest, confessions about environmental wrongdoing are emerging in online discussion groups because people with environmental shame do not just crave tips about what products to buy. They also want to deal with the emotional and existential fallout from their failures to live up to their own environmental ideals. Creating online communities or communicating with environmentally concerned peers offline to address these needs indicates that the solidarity, comfort, and advice from others are critical sources of relief for at least some people suffering from environmental guilt. Buying the latest green gadget cannot fulfill these needs.

Environmental marketing not only overemphasizes certain actions (buying goods or services) as the environmental thing to do but also focuses attention on a certain kind of agent: the individual. In doing so, it directs people away from group or collective solutions. In this paradigm, environmental protection is portrayed as hinging on the action of individuals, not nonprofits, governments, or the other groups or collectives. For instance, campaigns for hybrid cars often portray the solution to transportation pollution and energy use as dependent upon individual consumer choices rather than on community initiatives like increased access to safe bike paths, carpools, public transportation, or walkable infrastructure.

For decades, environmental marketers, including those using guilt and/or shame, have promoted individualism. Probably the most famous example is the Keep America Beautiful ad campaign of the 1970s which featured a man depicted as a Native American crying by a roadside full of litter. These commercials drew on people's emotions of guilt by relating their activity (littering) to the ugliness of a littered roadside and the tears of another. The ads were sponsored by Keep America Beautiful, which was portrayed as a grassroots activist group, but actually was one of the first "astroturf" movements as it was funded by the firms that invented disposable, nonreturnable cans and bottles and companies that used them. With this campaign, marketers deliberately attempted to shift guilt, shame, and responsibility to individual consumers by suggesting that *the* solution to the problem of littered roadsides was to change individual behavior to use trash cans instead of car windows (Beavan 2009, 65–6; Rogers 2005). Keep America Beautiful did not promote avoiding cans in the first place, instituting bottle return bills or policies taxing or banning the sale of disposable drink containers. All of those initiatives would require a group, if not collective effort, shift responsibility to the companies, challenge their business model, and ultimately threaten their profits.

The Keep America Beautiful campaign was one early initiative in a long and growing strand of environmentalism that presumes individual responsibility to be the root cause of environmental degradation and thus the necessary type of response. Norah A. MacKendrick found that articles in a Canadian newspaper from 1986 to 2006 about "the framing of chemical bioaccumulation" in human bodies increasingly used a framework of "individualization of risk, where individuals are encouraged to avoid contaminants through 'precautionary consumption' of green consumer goods" (MacKendrick 2010). In 2000, Patrick Novonty maintained that media coverage of how to respond to environmental issues consistently focuses upon activities for individuals within their households over offering suggestions about community activities or activism (Novonty 2000). The trend has continued to the present in most of the most famous environmental texts. In the book *An Inconvenient Truth*, Al Gore frames the section about taking action individually: "So here's what you personally can do to help solve the climate crisis." Fourteen pages are devoted to individual actions before the reader turns to three pages about collective and social actions, many of which are still focused on the individual agent (educate oneself) or on investing one's own retirement plan, before he finally suggests voting, "tak[ing] political action," and "support[ing] an environmental group" (Gore 2006, 305–21). In this book like others, when community-based or political actions are suggested, it is often at the end of a long list and with relatively little detail. The fact that individual, household actions are often easier to contemplate and act upon than joining a movement (especially given the modern western world's strong commitment to individualism) only reinforces this tendency.

The intentional emphasis of individualistic values and solutions to environmental problems is not merely a different environmental strategy but is often a form of greenwashing, typically considered deliberately "misleading consumers regarding the environmental practices of a company or the environmental benefits of a product or service" (TerraChoice Environmental Marketing 2007, 1) as cited in (Gibson 2009, 424).[8] I expand the term "greenwashing" to include misleading consumers about the array of options to address environmental problems and deflecting environmental responsibility away from companies, government, and other groups. Greenwashing may also include marketing that focuses consumers' attention on a relatively minor environmental problem while deliberately obfuscating larger issues in which a company is complicit. Campaigns to turn off lights in unoccupied rooms or use power strips to avoid the vampire power use of devices can direct attention away from the source of the electricity (e.g., coal, nuclear). Certainly, avoiding wasting electricity is important, but so is developing and encouraging clean energy through activism and regulation. As green marketing focuses on individual solutions, it distracts from the challenges of collective responsibility, guilt, or shame and the collective acts needed to address environmental challenges.

In this analysis of environmental marketing, I have shown how it unsurprisingly often serves the ends of selling products over achieving green ends. It can intentionally or unintentionally distract or deliberately manipulate people from the most environmentally effective use of their time and money, may promote goods and services that are broadly "green," or use tactics that unintentionally

[8] While such practices are common, "many companies are afraid to educate buyers about green options for fear of being called greenwashers" (Hopkins and Roche 2009, 87–8).

increase rather than decrease environmental harm. As it promotes individualism it can distract from more effective environmental activities and the complex realities of our multilayered forms of agency: individuals, memberships, and collectives. Finally, it may erode the dignity of agents by shaming them or by forestalling their participation in decision-making that affects them.

7.3 Environmental Shaming of Women, Especially Mothers: A Disproportionate Burden

Environmental marketing and interpersonal uses of environmental guilt and shame are not, however, evenly distributed throughout populations. Rather, they often disproportionally target certain people, particularly women, and especially mothers.[9] When targets of guilt and shame campaigns are already prone to experience guilt and shame due to social conditioning (as women often are), these disparities can have significant ramifications for the physical well-being, sense of self, and agency of the people in question. Besides misdiagnosing responsibility, such trends endanger the well-being of everyone. Furthermore, other agents (corporations, governments, men) are considered—and consider themselves—to be off the hook. To substantiate these claims, I will explore the gendered nature of environmental marketing, especially that which uses shaming techniques, identify cultural reasons why these strategies are particularly effective, and discuss the harms such strategies pose not only to women but also to all people and the environment at large.

On first glance, upper-middle-class people appear to be the most obvious target of environmental marketing, including those ads utilizing guilt and shame, as it is often assumed that they have the interest and, maybe more important from a marketer's point of view, the resources to buy products labeled as "green" (Atkinson 2014, 560; Smith 2010a, 72).[10] Targeting upper middle-class people for environmental guilt and shame campaigns also makes sense, as such people are significant contributors to environmental degradation and, relatively speaking, have ready power and means to make a difference, though this does not seem to be a major concern of many marketing campaigns for "green" household goods.

While socioeconomic status may be the most obvious presumed determinant of environmental marketing, more ads target women, particularly mothers. Mothers, especially working mothers, who may have guilt about working are

[9] I recognize the limitations of the binary notion of gender roles and expressions used in this section. The traditional gender stereotypes which dominate environmental marketing—and limitations of research on gender and the environment and my primary source data about environmental guilt and shame—hinder a more detailed analysis.

[10] Of course, many socioeconomic groups have resources to engage in some environmentally friendly practices such as turning off the lights when leaving a room.

believed to purchase products to assuage their guilt (Coulter and Pinto 1995, 699). Long-held stereotypes about women often ground such campaigns, e.g., women are more emotional; associated with nature whether life-giving, wild, unpredictable, or dangerous; and presumed to be nurturers, caretakers, and responsible for homemaking including household purchases.[11]

As we consider shame in particular, the associations with women only increase. In *Hiding from Humanity*, Martha Nussbaum argues that the societal shaming of individuals and groups is a ubiquitous part of human life arising from the ways people experience themselves and interact with each other. As people experience their limits, they want to feel powerful and to think of their limits as insignificant. To achieve this, Nussbaum maintains that people set up "in" and "out" groups where they are "normal" and others "abnormal." People classify others as abnormal according to some standard, often related to gender, sexual orientation, or ability (Nussbaum 2004, 173–4). Women and GLBTQI+ folks have often been "othered," or shamed. Similarly, shame and poverty often go hand in hand (Chase, Kyomuhendo, and Bantebya 2015).

Many feminist scholars note that shame is often culturally associated with power, since it may be induced by judgements the powerful make about the less powerful (Eurich-Rascoe 1997; Locke 2007, 148), as seen in the work of such influential thinkers as Aristotle and Freud (Manion 2003, 22). Shame's connection to bodies, particularly women's bodies, which may be perceived as physically weaker than men, messy or out of one's control (i.e. in menstruation, pregnancy, or menopause), or associated with shameful sex (i.e. as a sexual being or a victim of sexual violence), reinforces this power differential in women's experience (Dolezal 2015; Kaplan 2002; Taylor and Wallace 2012, 77). Additionally, Sandra Lee Bartky maintains that women often face "chronic" shame as they experience ongoing social pressure and internal judgment about their bodies, capacities, and selves (Bartky 1990, 85–96; Lehtinen 1998). Transitory experiences of environmental shame can exacerbate persistent shame, since these moments seem to corroborate the experiences and expectations of the shameful. Simultaneously, chronic shame may prime women for feeling it ever more intensely, as it causes them to have a negative or diminished sense of self.[12] Thus, there are many reasons why marketers of environmentally friendly household products would target

[11] These tropes are well studied, so I limit my summary (Merchant 1980; Smith 2010a). Research often indicates that women are more likely to express concern for environmental issues or nature, though they may be less likely to take public and more likely to take private action about them (Davidson and Freudenburg 1996; McCright and Xiao 2014; Somma and Tolleson-Rinehart 1997; Yates et al. 2015). Such correlations do not imply biological differences rather than cultural conditioning.

[12] Other demographic groups including racial and ethnic minorities and GLBTQ+ people may also experience chronic shame (Hequembourg and Dearing 2013; Macdonald, Abbott, and Jenkins 2012; Saraswati 2012) and thus be more susceptible to particular episodes of shame, including environmental shame.

women, why those using shame-based tactics would specifically target women, or why environmental marketers targeting women would use shame-based tactics.

In fact, gendered marketing, particularly using guilt and shame, about issues that would now be considered environmental, predates the modern environmental movement. In the 1920s, as the toxicity of lead, especially to children, became known and other nations were banning its use in paint, Dutch Boy and National Lead paint campaigns encouraged the use of lead paint in the United States. These ads were aimed at people involved in decision-making about paint, especially "master painters, homeowners, wives, and mothers" (Markowitz and Rosner 2000, 39). Marketing campaigns often used images of babies or young children, claiming that lead paint was the "purist" or "safest" option and suggesting that using anything else might put your child at risk (Markowitz and Rosner 2000, 40–41). Later, when U.S. American companies started to admit the dangers of lead, they blamed parents, especially mothers, particularly of low-income, inner-city neighborhoods, arguing that if they would just keep house better or be a better mother their children would be fine (Bellinger and Bellinger 2006, 855; Markowitz and Rosner 2000, 43; Rosner and Markowitz 2008). Such judgments and the resulting social policies often induced guilt and shame in mothers for failing to prevent their children from being exposed to, and often poisoned by, lead (Bellinger and Bellinger 2006, 855; Markowitz and Rosner 2000; Rosner and Markowitz 2008).

Indeed, the expectation that the health of children is up to and in the control of mothers is a common trope (Taylor and Wallace 2012, 82). Some potential mothers spend a year or more trying to purify their homes and bodies before even trying to get pregnant (Lindsey 2012)—a task at which they will never fully succeed at, given human physiology and the prevalence of toxins in our world. Many ads suggest that one must buy organic food and sustainable toys and use non-toxic, "natural" cleaners to keep one's family safe. Pregnant women and mothers of infants are particularly targeted for such campaigns. In Lucy Atkinson's study of green ads in *FitPregnancy* she found that ads typically prioritize products for and the experience of the child over the mother, often eliminating her fully or partly from the ad. Ads did this by focusing photos on the baby in a mother's arms, eliminating the mother's face and all of her but for a bit of torso and arm, or choosing images of babies alone rather than with a mother, and "encourage[ing mothers] to [. . .] leave as little evidence of her presence as she can" on the earth (Atkinson 2014, 562–5). Atkinson maintains that mothers' experience and knowledge are ignored or supplanted as the brands are portrayed as the ultimate experts on raising a healthy, "green" baby (Atkinson 2014, 565–7). All of these trends together, she argues, push a self-sacrificing, self-eliminating view of motherhood (567). To do otherwise is to risk one's child, and, thus, to be less than ideal (i.e. to be shameful).

Corporate marketing and media are not, however, the only forces involved in inducing and experiencing environmental guilt or shame among women.

Interpersonal relationships can also play a role. Robyn Harding's neighbors (discussed in Chapter 5) explicitly and implicitly influenced her ideals and made her recognize that she was not living up to them. She became aware of and adopted new environmental priorities such as recycling when she moved to new cities where such actions were valued and supported by infrastructure. While Harding rarely describes friends and neighbors overtly shaming her, she is keenly aware of the pressure to fit in and wonders about her identity and worth in relationship to others, others' views of her, and existence in general (Harding 2009, 44–5).

Similarly, the rise of social media, with its "curated" photos and stories often only including the most harmonious, successful, attractive, and apparently environmentally friendly activities in beautifully shot photographs, can exacerbate guilt and shame about one's life including environmental behavior. A steady diet of such selective news and related environmental how-to books can significantly increase the pressure women feel to have perfect home lives (Schoppe-Sullivan 2016; Schoppe-Sullivan et al. 2016), given stereotypical social expectations about who takes care of homes and who is socialized to care what others think about their relationships and homemaking. In this context, it may be especially meaningful when authors of environmental or homemaking blogs, who often serve as role models to their readers, confess environmental guilt or shame online. In doing so, they (even if unintentionally) model expected and accepted emotional reactions and behavior for their readers. Thus, they normalize environmental values and the guilt and shame springing from them, particularly for the largely female readership of homemaking media. All of these cultural trends set up women to be disproportionately shamed and feel more environmental shame.

Targeting women for environmental shame experiences and the disproportionate experience of environmental shame by women are problematic for a number of reasons. For women already experiencing chronic shame, internal and external expectations about being an environmentalist and shame about not meeting these ideals can be a self-perpetuating cycle. Environmental expectations are often incredibly, unattainably high. Women will never be able to be perfect environmentalists because humans are not perfect, and even if perfection was possible, environmental scenarios require the balancing of multiple priorities in complex, dynamic situations, often with incomplete information. Even if women could optimize every environmental decision for their household, they would not eliminate anthropogenic environmental disruption or fully protect their families, given their enmeshment in physical, political, and social systems.

Intensifying the unrealistic nature of environmental perfectionism is the possibility that shame can drive one toward narcissism. The shameful may focus on their own emotions and self-concept so much that they face "paralyzing distress" and neglect other, potentially larger, concerns. According to political scientist Jill Locke,

> One of shame's most poisonous consequences is the way in which it overwhelms the subject so that she is unable to think beyond herself. Rather than focus on changing the world in ways that might lessen her shame, the shamed subject focuses on changing herself so that she might accommodate the demands of her milieu. (Locke 2007, 151).

In environmental terms, this can manifest in frequent online confessions or in overemphasizing daily household actions, such as bringing reusable bags to the store, rather than reexamining habits of flying several times a year, which has an environmental impact several orders of magnitude higher than any decision about grocery bags. Changing laws and infrastructure (assuming the laws are enforced with significant consequences and are constructed to avoid further harm to the vulnerable) could have a larger effect than individual changes to daily habits, because legal changes can change the actions of even those without the desire, time, or funds to prioritize the environment (Jacquet 2015a, 7). Without systematic change, the self-centeredness of shame is likely to indirectly exacerbate shame because people will continually experience their own fallibility and systems they cannot change with small isolated actions.

Compounding the links between environmental shame and women is the fact that the environmental movement has often, especially in recent years, been portrayed as a movement of many individuals changing their household activities to yield a large positive cumulative effect on the environment. Thus, people taking care of the home, or those assumed to do so, bear the brunt of environmental decision-making and implementation. In the United States, these people are typically women. Women in heterosexual couples in the States spend more time on household tasks even when also working outside of the home (Hochschild 1990) and even as men increasingly participate in household chores (Altintas and Sullivan 2016, 458). Though single men and women experience less temporal disparity regarding housework than their heterosexually married or partnered neighbors (University of Michigan Institute for Social Research 2008, 3), the persistence of this disparity suggests that women face more pressure to maintain their households regardless of relationship status. Advertisements and media aimed at women's housekeeping activities including shopping for the home (Smith 2010a, 70). Media stereotypes of messy bachelor pads likely reflect and reinforce such trends.

Little academic attention has been given to the relationship between expectations about household labor and environmentalism (Smith 2010a; Vinz 2009). The available studies indicate that women bear the burden of environmentally related household tasks such as recycling (Oates and McDonald 2006), and may be subject to increased pressure to be environmentalists through advertising, especially if they have children (Atkinson 2014). In a household striving to be environmentally friendly, the amount of time spent gathering information and making decisions about household activities can rise exponentially. Once such decisions are made, the extra time and effort spent to implement the decisions—to find the desired products, cook from scratch, hang clothes to dry, wash reusable containers, and mend slightly worn clothes likely also falls to women, given the household time-use data discussed above. Thus, increased green expectations in the general population are likely to lead to disproportionately increased expectations for women and disproportionate time spent on green household tasks—conditions that set women up to experience more guilt and shame, especially if they are already more concerned with caring for their families and the environment.

Stereotypical portrayals of women's agency as less than that of men's in the media, especially in direct marketing, intensify the environmental shame of women. In the aforementioned study about the dominant trope of green consumerism in women's magazine articles about the environment, the only two articles without a "green commodity component at all" were those written by men about the environmental actions of men. Smith concludes that these articles "contained an insidious subtext: women either support the environmental agency of men or act as an obstacle to the environmentalist goals of men [...] the possibility of meaningful feminine agency is omitted" (Smith 2010a, 77). With such portrayals of women's agency (or lack thereof), even in places that one might expect to see women being empowered such as women's magazines, we find one more factor reinforcing chronic shame of women in the form of a diminished or limited sense of self.

The negative consequences of environmental shame, are not, however, confined to the particular women experiencing them. Insofar as women experience more environmental guilt and shame than men, they will likely feel, take, and be expected to have more environmental responsibility—an unrealistic burden for them and a situation that, comparatively speaking, can let men off the hook, especially as men have often made governmental and business decisions with major environmental impacts. Additionally, if men do feel or judge themselves to be guilty, they may feel social pressure to be "masculine" in opposition to the femininity often associated with environmentalism (Bennett and Williams 2011, 52–4).[13] This state of affairs is ethically problematic because it unjustly assigns responsibility to some people and unjustly ignores others' responsibility, all while making it less likely that effective action occurs. After all, environmental shame has the potential to exacerbate the physical, economic, social, and psychic effects of environmental degradation, including climate change, by obstructing or

[13] Several studies indicate that the gendered expectation of and vision of environmentalism is weakened and may be indiscernible among young people (Bennett and Williams 2011, 72; Suplico 2009), so this stifling of men's responsibility may be becoming less of a concern, even if women's disproportionate self and community burden for the environment persists.

postponing significant action. This may occur through household activities including refusing to recycle or failing to conserve energy or water. It can also cause one to deny that there is a problem at all, leading to apathy or opposition to environmental movements or regulation.[14] Such denial and obstruction is, of course, important when human civilization faces a dire threat such as climate change which requires a global response, including the developed nations that have contributed most to the problem. These nations need to be a significant part of the solution in order to right past wrongs through apologies, financial and/or physical restitution, respect for those harmed, and new laws to prevent reoccurrences of harm; to set an example for others; and to contribute technical and political expertise to the cause. Additionally, unless the developed world changes its ways, slowing climate change will be impossible.[15]

Finally, environmental shame can be particularly problematic because the women (and men) feeling it are some of the people who already have environmental values and have some of the most wealth, education, and power on the world's stage, even if they are relatively disempowered in their own nation or home. If they do not act, mitigating and adapting to climate change and other global environmental threats will be difficult, if not impossible. As their inaction continues and the physical problems of climate change worsen, the experience of shame may increase, driving a positive feedback loop of shame and environmental degradation.

7.4 Naming and Shaming

Another popular intentional use of shame to elicit behavioral change targets memberships and collectives. In these naming-and-shaming campaigns, nonprofit groups and grassroots coalitions typically name companies or governments as shameful. While such campaigns can avoid some problems of individualism and power imbalances and can provoke significant behavioral changes, they are not without risks to shamers, shamees, and other agents.

Insofar as naming and shaming is about an agent's character and not just bad actions, its use of the term "shame" aligns with my use of the term. Yet, naming and shaming does not usually intend to lead to totalizing shame, where the entity in question is deemed beyond the pale for all time. Rather, naming and shaming often assumes that its force will provoke action to avoid or limit reputational damage; indeed, shamers often specify how the shamed should change to follow

[14] Certainly, shame does not motivate all climate deniers, but anticipatory shame may affect some. If they did accept that climate change was real, this reality would devastate their vision of humanity. For similar ideas, though not explicitly about shame, see (Moe-Lobeda 2013, 96; Northcott 2013).

[15] Given the rising rates of emissions by the developing world, the developed world cannot address the climate crisis by itself.

the norm. For instance, recent campaigns have tried to get banks to stop lending money to fossil fuel projects because of the role fossil fuel consumption plays in climate change. Such targeted actions may indicate that such campaigns may be more about action-oriented guilt and the reputation that accompanies it than about whole-self-judging shame.

Naming-and-shaming campaigns have a number of advantages. First and foremost, they are often understood to be quite effective at changing behavior. In a 2002 study, Mark Stephen investigated environmental disclosure programs, a popular technique of governments and nonprofits to decrease environmental degradation (Cisneros et al. 2015, 4; Johnson 2011; Stephan 2002, 198; Tian et al. 2016; van Erp 2014, 3210). Stephen reviewed several theoretical reasons for collecting and disseminating such information at local, state, and national levels as well as the available empirical evidence to support or undermine the theory. He finds that "the strongest evidence" for the shame theory "comes from the literature on the connection between stock prices and environmental performance," which indicate that environmental news shapes markets and that "drops in stock prices have led to attempts by affected firms to reduce emissions (or, at least, their on-site releases)" (Stephan 2002, 198). Few studies to that point had examined the relationship of citizen action and corporate action, though some indicated a relationship between corporate behavior and communication with the public. Since then, additional studies have begun to fill in this gap in the literature. For instance, Rachel Stewart documented "one of the most successful examples" of "public shaming," that which arose in response to the documentary *Blackfish* about orcas at SeaWorld (Stewart 2016, 725). Stewart notes that though the film probably had an educational, not shaming, aim, shaming still resulted. As the public considered SeaWorld's actions depicted in the documentary as shameful, "[a]ttendance at parks dropped, stocks fell, and earnings slumped" (Stewart 2016, 726). A longstanding partnership with Southwest Airlines was severed after petitions. Several movies took scenes at SeaWorld out of films before releasing them. SeaWorld responded to the pressure by discontinuing captive breeding programs and deciding to "stop hosting orca shows by 2019" (Stewart 2016, 726). Proponents of naming and shaming point out that such significant results are often achieved with relatively few financial costs compared to developing and implementing new laws or other methods of changing environmental practices (Cisneros et al. 2015, 20; Jacquet 2015a, 13, 24, 97, 172–83). (Of course, such considerations are rarely based on a comprehensive study of all the effects of naming and shaming discussed below.)

Naming-and-shaming campaigns can also be advantageous in that, if constructed well, they can target the few agents who disproportionally contribute to a specific type of environmental degradation and/or those whose behavior must be changed to enable wide-scale change to occur. As Jennifer Jacquet points out, it is often the case that a few agents disproportionately contribute to environmental harms: "Just ninety corporations [...] are responsible for nearly two-thirds of historic carbon dioxide and methane emissions" (Jacquet 2015a, 66). Thus, she argues that if these outsized contributors to climate change are shamed into behavioral change, future emissions may significantly decrease. Without such major players altering their actions, the behavioral changes of individuals will have a limited effect.

Decreasing environmental disruption is not enough to deem naming and shaming ethically advantageous by itself; the means by which a good end is achieved matter, as does the end itself. Thus, it is significant that the process of naming and shaming supports many values besides physically effective environmental change. For instance, "naming" may help stakeholders to have more information about a situation than if they were only relying on self-disclosures by the corporation or nation being shamed. Naming and shaming may also draw attention to and foster a broader range of values such as intrinsic value, the right to clean air and water, or the ability to participate in traditional place-based rituals. Consequently, these campaigns could open discourse to a fuller consideration of the goods of human life than narrow economic considerations of the powerful. Facilitating the existence and development of such values may demonstrate respect for the dignity of diverse participants as their values influence the conversation. Dignity of all people is also recognized when naming-and-shaming campaigns center on human rights. If the campaigns also acknowledge consideration of nonhumans, recognition is expanded.

When naming and shaming relies on the general public rising up against a corporation or nation, a broader sense of participatory justice is fostered than during most of the marketing strategies discussed in the previous section. Indeed, naming-and-shaming campaigns, particularly those led by nongovernmental organizations against corporations, often avoid the problems of power differentials that occur during corporate marketing campaigns employing guilt or shame. Naming-and-shaming campaigns are often led by (or on behalf of) marginalized people, communities, and nonhumans with little capital or formal political power against major corporations with teams of lawyers and media consultants, vast financial resources, and legislative connections. While power is certainly exerted during naming-and-shaming campaigns, the broad participatory nature of this power, likely for richly diverse ends rather than narrow ends of selling products, makes naming and shaming of corporations or governments more palatable than that of a few organizations misleading and greenwashing the masses. Of course, there is no guarantee that the masses using naming-and-shaming techniques will be virtuous. Naming and shaming will, however, be more likely to be aligned with participatory justice and support the values of a variety of participants than environmental marketing campaigns in which a corporation attempts to change people's behaviors for its own benefit and holds so much power that it can effectively overshadow other values and actions.

Particular relationships between shamer and shamee may entail that those being shamed have obligations to pay attention to the concerns of shamers, a condition that mitigates the manipulative possibilities of shaming. Since democratically elected governments are supposed to represent and be accountable to their citizens, citizens should have the right to influence their government through a variety of mechanisms including shaming when they feel the government does not represent their values. (The variety of values in any one nation certainly complicates such accountability.) Relationships to corporations are more complex. Many people think that corporations have particular responsibilities to their shareholders such that shareholders can pressure companies to change their social policies. While shareholder action sometimes does work, corporations are legally bound (at least in the U.S.) to make money for their shareholders. If enough people share the values of the shamers, changing company policy may be necessary to maintain a consumer base or avoid new regulations, but if the values are not widely held or are deemed too costly to follow (particularly under the short timescales typical of such assessments), the corporation (aside from B-corporations which are required to consider profits and social effects in their decision-making) may not be able to act accordingly. Yet, due to the way actions and associations can shape one's identity (as discussed in Chapter 4), shareholders are right to be concerned when corporations in which they invest act contrary to their values. Additionally, consumers may use shame to influence the behaviors of corporations whose products and services they consume. Such efforts are likely to be most effective when they align with the corporations' bottom line, whether directly, as when improving energy efficiency of manufacturing cuts costs, or indirectly, as when naming and shaming comes with significant reputational harms that influence purchasing behavior or business partnerships. In each of these cases, the relationship between shamed and shamer shapes the character, outcomes, and acceptability of the shaming.

Naming-and-shaming campaigns are not, however, without limits and risks. Such campaigns can be employed to discredit others based on hyperbole and innuendo rather than with faithful representations of the shamee's actions. They can be used to prop up selfish or short-sighted values. As with attempts to induce shame in individuals, collectives experiencing shame may deny the problem or become paralyzed in the face of their shame. Shame, humiliation, or stigmatization may also lead targets of shaming campaigns to be defiant of the norms on which the shaming is based (van Erp 2014, 3212). Shamers, particularly if they try to shame based on values that are not widely held or if they try to shame powerful opponents, may incite pushback, anger, or retaliation. For instance, shamers may be the subject of smear campaigns to undermine their information, arguments, values, and credibility. Shaming activities may trigger new classes of legal restrictions or punishments, as when a new law prohibited videotaping and publicizing factory farm activities (Jacquet 2015a, 152). Such laws may make it more difficult for the shamer's values to be expressed in the future.

Furthermore, successful shaming a corporation or industry can have many unforeseen consequences. Businesses may close or move overseas. Beyond the local social and economic consequences of such moves, environmental and social consequences of the business may be relocated rather than eliminated. When the new national host of a polluting corporation has lax social or environmental laws, the net effect of shaming may be more rather than less problematic behavior. Shaming campaigns, then, need to be carefully constructed to provide pathways to holistically constructive solutions.

Shaming in general, and maybe particularly shaming the powerful, may also pose moral dangers to the shamer. Shaming may be motivated by the desire to cut another down to size in comparison to oneself (Ronson 2015, 88–9; Thomason 2018, 197). Shaming also involves moral risks including "being self-congratulatory and morally naïve," forgetting the other humans involved (Thomason 2018, 197), and being "brutal" to others in a way that can change one's habits, character, and identity, especially if the shaming occurs online, where one can disassociate from or even for it and thus be unlikely to be held account for it (Ronson 2015, 88–9).

Even when naming-and-shaming campaigns have positive outcomes, their results may be far from sufficient. To have a successful campaign, targets may be chosen based on the likelihood that they will change in response to shaming rather than the agents who contribute most to the problem. For instance, Michael John Bloomfield notes that the "No Dirty Gold" campaign focused its shaming initiatives on the United States, though both India and China use more gold, a situation that continues a disproportionate focus on the United States and may inhibit possibilities of global change (Bloomfield 2014). On the other hand, global naming-and-shaming campaigns may preclude local values and autonomy (Bloomfield 2014, 269–70).

Naming-and-shaming campaigns, like environmental marketing campaigns, may also promote individual, consumeristic solutions to environmental problems (Bloomfield 2014, 267–8) by aiming to label particular products as "green," "environmentally friendly," or "dolphin-safe" rather than comprehensively changing whole industries or systems. As Jennifer Jacquet argues, the end result of the 1980s initiative to shame tuna companies was not an industry-wide ban on fishing practices that incidentally kill dolphins but rather a product certification indicating that labeled brands of tuna are friendly to dolphins (Jacquet 2015a, 3–9). While labeling has saved dolphins, the solution transformed dolphin safety into an issue of individual consumer responsibility instead of systematic change. Indeed, she argues that solutions based on individual feelings of guilt has been all too often offered as *the* solution (5).

Further risks of naming and shaming, as well as a few benefits, become apparent when we explore the differences between memberships and collectives experiencing guilt and/or shame. These risks are relatively unexplored given the dominance of individuals in ethical analysis and insufficient theorization of collectives, groups, and their interaction with individuals. For instance, both Jacquet and Nussbaum find shaming groups more acceptable than shaming individuals because of their presuppositions about individuals and groups (Jacquet 2015a, 24, 174; Nussbaum 2004, 211–13, 241). Yet, as we shall see, even if one is only concerned about individuals, there are good reasons to be wary of shaming groups and collectives. If one is concerned about collectives as well, the reasons only multiply.

Recall that membership guilt and shame are based on the judgments and/or feelings that arise as one reflects on the actions and/or existence of a group to which one belongs or of other individuals in that group. For example, as I first drafted this section in February 2017, members of the U.S. legislature were working to confirm a new leader for the Environmental Protection Agency who planned to gut the institution. Some U.S. Americans may have experienced membership guilt around these events because their fellow citizens were committing actions that they found reprehensible. They may have also experienced membership shame based on the recognition that other members of one's group are the kind of individuals who can make such decisions. Additionally, membership guilt and/or shame may be elicited indirectly in an individual when a group the individual is associated with is the target of a naming-and-shaming campaign. For example, protests against bank-funded pipelines in front of banks could cause bystanders to recognize the connection between their banking habits and climate change. Note that in all of these cases, the membership guilt or shame is something assessed by an individual. It is not an assessment or feeling shared collectively or perceived to be a collective act. One sees the others in one's group acting as individual agents in ways that influences oneself as an individual.

Pointing out this possibility is an important corrective to many advocates of naming and shaming, including Jacquet. She does not consider the possibility that individuals may be unintentionally shamed due to their association with named and shamed corporations or nations (Jacquet 2015a, 4, 7). Similarly, Nussbaum finds the shaming of groups less worrisome than shaming individuals. In *Political Emotions*, she claims that "shame is particularly likely to be healthy when it is collective" (Nussbaum 2013, 364). In her one extended analysis of the positive possibility of shaming a group, Barbara Ehrenreich's shaming of U.S. Americans for their inattention to and inaction regarding the working poor, Nussbaum favors Ehrenreich's shaming of U.S. Americans in part because it is informal and diffuse (Nussbaum 2004, 211–12, 236). Informal shaming prevents the state itself from going against the liberal values that Nussbaum prizes. Certainly, avoiding conditions where the state will undermine its own values is important if one supports those values and consistency in the application of values. But informal shaming does not necessarily limit the number of people under social control, since social control is not limited to formal judiciary systems or state power. Individuals, nonprofits, and other collectives can also induce shame in each other. Nussbaum finds Ehrenreich's shaming helpfully diffuse because it is directed to U.S. Americans and U.S. American society in general rather than singling out any one individual as responsible for the systemic challenges faced by the working poor. Nussbaum has no apparent concern that the self will be degraded by the group's shame because she reserves identity for individuals—collectives do not (quite) have it. Since, for Nussbaum, collectives do not have a distinct identity more than the sum of their parts, or (more importantly) dignity of their own, there is no danger in shaming groups for Nussbaum. She finds that individuals are only partly implicated in the actions and status of the group (membership guilt and shame) and therefore individual identity is not tainted by that of one's group.[16] In Nussbaum's analysis of Ehrenreich's work, the group in question being shamed—all of America—is large enough that it precludes singling out individuals as distinct from and abnormal compared to the group or people in power. The diffuse nature of Ehrenreich's shaming is also likely palatable for Nussbaum, since Ehrenreich implicates herself in the shame (Nussbaum 2004, 244), further avoiding the bifurcation of normal and abnormal which disrespects and degrades another or a group of others, as Nussbaum argues throughout *Hiding from Humanity*. By avoiding this dichotomy, Nussbaum seems to think that Ehrenreich can avoid the devastation of shame.

Limiting the division between shamer and shamee can be helpful to avoid shame's destructive tendencies. To the degree that one's identity is kept separate from the wrongdoers, one is less likely to yield shame and paralysis because of what the members of one's group does (Van Vugt 2009). Such a separation may enable individuals to have the desire and courage to call out their own group members or fight to change the way the group acts (e.g. educating their associates, calling elected officials, protesting, or working to elect like-minded individuals). Yet, individual experiences of guilt and shame feelings or assessments of guilt and shame states due to one's membership in a group can have the same implications as those experienced by individuals about their own acts as

[16] While Nussbaum does not explicitly say one cannot take on the shame of one's collective, her prioritization of individual agency and avoidance of ideas of the metaphysical contamination of self suggest this interpretation (Nussbaum 2004, 14). She does recognize that particular classes of people (women, disabled, homosexuals, etc.) have often been shamed, but her accounts of such shaming rely on each individual having the characteristic of the class rather than catching the shame of others. She also notes that public shaming of large corporations is an example of this positive and permissible shame because corporations have no ability to feel lack of dignity and do not have dignity (244). She, however, quickly returns to her focus on the legal system and state shaming of individuals, not quite coming to a clear conclusion about whether the state could experience this type of shaming (245). Thus, she seems to consider shame of memberships but not collectives as I define them.

individuals—reparative actions in the case of guilt and stigmatization, hiding, denial, and shame spirals in the case of shame (van Erp 2014, 3212). Shaming a larger unit including oneself will not necessarily avoid these problems of shaming.

Membership guilt and shame may also heighten the negative consequences of individual guilt and shame due to the cumulative effects of the membership group's actions which may yield results greater than the sum of individual consequences. For instance, the greenhouse gas emissions of any one person may have a negligible effect on climate, but together some groups (e.g. U.S. Americans) have a monumental effect. Recognizing this, an individual may experience more guilt or shame than they would otherwise. The complexities of climate change, including its systemic nature and multiple temporal and spatial scales, only adds to the challenge.

Yet, the experience of collective shame, including that arising from naming and shaming, may be more devastating than that of individual or even membership levels. While some everyday environmentalists are recognizing and responding to collective identity, responsibility, guilt, and shame as discussed in Chapters 2 and 5, these are often implicit, nascent moves lacking rich conceptual or cultural infrastructure. The strong individualistic nature of most modern, Western worldviews hinders the development of such infrastructure. Collective assumptions and values may be so pervasive that it is difficult to imagine an alternative to an action or state of existence, let alone try to implement it. When we consider systemic problems such as climate change through the lens of collectives rather than individuals, the numerous, overlapping, and often diffuse collectives implicated alongside individuals and membership groups become apparent. If an agent is already overwhelmed by the scope and severity of climate change, these layers of agency and responsibility will likely challenge action inspired by guilt and may exacerbate the paralysis of shame, particularly since a much clearer sense of the interaction of and relative weights of different kinds of agency is needed than I have been able to articulate here. Compounding these challenges is the fact that many collectives involved in environmental degradation are not intentionally constituted collectives, so taking steps to evaluate and change their actions requires an explicit recognition of the collective (Isaacs 2011, 144–54). Without the necessary conceptual and infrastructural frameworks, dealing with guilt and shame will be even more difficult than it already is.

These challenges can also inhibit attempts to deal with shame. Recall that everyday environmentalists responding to online confessions frequently suggest or model reparative actions, and contextualize the identity of the agent by recognizing at least partial goodness of the agent as evidenced by the good things the agent does and thus the goodness of at least part of the agent's identity. Thus, they enable a move from whole-self-judging shame to action-oriented guilt or at least to partial-self-judging shame mixed with guilt. Similar processes may happen with collectives, as when a nonprofit calls out a corporation for its environmental track record and is willing to stick with the corporation as it works to repair its actions and deal with its shame. Here the nonprofit thinks the corporation can change, that its identity is not wholly defined by the environmental degradation. Yet, the nature of collectives and the scale and diffuse nature of some of them may confound such processes. For instance, it is not clear to everyone whether collectives are the kind of being that can be in a state of shame, be in a praiseworthy state, or have a mixed identity. My argument in Chapters 3 and 4 suggests that they can, but such ideas are far from dominant in our individualistic society. Much more work is needed to clarify these ideas and determine if and how they can be influential outside the academic realm.

Insofar as collectives are larger and may even represent or be equated with humanity in general, experiencing collective shame could be even more devastating than personal shame because it could be seen as indication not only of one's own identity but also of one's collective or human nature as a whole. When radically shameful notions of large groups or humanity are accepted, as we see in some rhetoric about climate shame and in some Anthropocene literature in general, it can be difficult if not impossible to imagine that individuals or collectives could do or be anything other than shameful. When dealing with climate change in particular, the scope of the horror (hundreds of millions of climate refugees predicted, species extinction on a vast level, threating human civilization as such) raises further questions. Is anything other than a totalizing negative judgment for those deemed responsible appropriate? Is there any differentiation within the character of collectives responsible for climate change, or are they wholly bad? The temptation is to judge them as being beyond the pale, as we see in many activist attempts to shame fossil fuel companies. With such whole-agent-judging shame is there any hope of remediation or change? Do they have intrinsic value or dignity such that we should be concerned about the harm shaming causes to the collective as collective, in addition to any harm that shaming it may cause to individuals? Significant additional research about the metaphysics and axiology of collectives is needed to address these questions. It is, however, beyond the scope of this project.

Even if one's concern is only for individuals, shaming collectives may have disastrous results. Recall Jaspers' argument that the shaming of Germans after World War I was a factor contributing to World War II. What worse shaming could happen if an agent is judged a prime contributor to climate change that is changing planetary systems to be less hospitable for humans? What punishment? What worse reaction by the shamed? Are we already seeing this reaction among collectives that refuse to face their culpability for climate change and what it says about their identity? Do such potential dangers suggest that environmental shame should never be used in relation to climate change? More research is needed on these issues, especially with respect to the possibilities of shaming fossil fuel corporations and the United States.

Possibly most devastatingly, if a collective responsible for climate change is a large diffuse collective such as humanity as a whole and shame is involved, there is no easy interlocutor who can contextualize the identity of the collective and point to a more nuanced vision of its character or the possibilities of future change. An individual or smaller collective in this supercollective is likely to find it difficult to envision possibilities disentangled from the collective. Conceptually, an alternative metaphysical entity such as God could provide such feedback and/or possibilities, but such a notion is, of course, conceptually challenging, particularly in pluralistic and often secular contexts. There are no readily accessible, existing well-functioning systems of relationship, accountability, or moral support to deal with the shame of many collectives. No one to (as a peer or equal entity) point out the variations in the collectives' actions and identity, no one to model good behavior, no one to participate in transformative acts with the collective. Even for collectives smaller than the supercollective of all humans such as a fossil fuel corporation, systems of accountability are difficult to come by. Naming-and-shaming initiatives attempt to fill this gap, but often do not have long histories with those being shamed such that the accuser will be believed or trusted or such there is a shared set of values. Similarly, relationships of naming-and-shaming campaigns may not project far enough into the future or be comprehensive enough to deal with the many ramifications of the shameful activity and being.

In sum, collective shame is more complicated and more risky than individual and membership shame, not just because of the potential scales involved when considering the largest collectives but also because there are dangers particular to collectivity and because it more difficult to conceptually and practically move away from shame's totalizing judgment of identity at collective levels. Thus, even as naming and shaming collectives can effectively catalyze behavioral change, enable agents to clarify their ethical priorities, and encourage others to adopt and live out such values, it is far from a clear answer to environmental challenges as framed by many of its advocates.

* * *

As I take stock of the ethics of intentionally shaming identified in this chapter, several considerations loom large: concern for the dignity of persons such that their worth ought not be degraded by shaming; the manipulation possible when inducing guilt or shame; disproportionate burdening of individuals and certain demographic groups, especially women and even more particularly mothers while relatively letting men and collectives off the hook; and concerns for effectiveness, whether Jacquet's more narrow concern for cheap, quick, and effective responses to climate change or my more comprehensive attention to shame's consequences for individuals, collectives, and their interaction. These priorities likely have a similar root concern: the dignity of and respect for entities in and of themselves. Without this presupposition, manipulation of people through environmental marketing or the disproportionate shaming of women would not be a problem. Similarly, concern for effective solutions to urgent challenges such as climate change is likely grounded at least to some degree on the dignity of entities (individual or collective humans, other biota, ecosystems). Of course, one could value effectiveness in and of itself, but—at least in literature about environmental shaming, such as Jacquet's—the prioritization of effectiveness seems more concerned with having a good outcome for people, a priority that presumes their value and dignity.

While dignity can link the various ethical considerations of inducing guilt and shame and connects them to shame with its concern for identity and worth, collapsing the ethics of guilt and shame to dignity alone would be insufficient. Most studies of dignity are so narrow that using this term alone may distract one from the interaction of individuals and collectives or from broad concerns of effectiveness. All of these ethical ideas are important.

This complex ethical picture precludes a definitive yes or no answer to the question "Should anyone induce guilt or shame?" As indicated above, inducing these moral emotions can prompt significant behavioral change and reflection on values, but at a significant cost to the identity of agents and all affected by their actions. Additionally, recall that ethical development occurs in part through the experience of moral emotions. Inducing guilt and shame and experiencing guilt and shame, regardless of their source, seem so fundamental to being a human that it is difficult to imagine human life without them. If we reject the use of guilt and shame as some critics such as Nussbaum seem to want to do, then we run the risk of alienating our ethical approaches from the realities of human life. Even if one just wanted to reject guilt and shame as strategies due to their ethical risks, one would need to deal with the fact that others will almost certainly unintentionally experience such feelings. Thus, the question of whether we should induce guilt or shame is intimately linked to the question of how we can deal with the fact that such feelings will inevitably occur. The next chapter will identify the conditions necessary to enable agents to deal with consequences of guilt and shame.

8

The Ethics of Inducing and Responding to Guilt and Shame

The many possible negative consequences of environmental guilt and shame and the fact that guilt and shame seem to be such fundamental aspects of human experience that we cannot avoid them, indicate that people need ways of dealing with these moral emotions. Furthermore, if, despite the many dangers of deliberately inducing them, agents decide to do so, they will need guidelines to ensure that their actions have the desired effects while diminishing unintended consequences. Although everyday environmentalists have been implicitly developing ways of dealing with environmental guilt and shame, particularly through confession rituals, I will not recommend relying on spontaneous ritualization as a sufficient way to respond to guilt and shame. The threats are too significant. Rather, in this chapter, I rely on material from previous chapters to identify conditions needed in the shamer, shamee, their relationship, the broader community, and the physical system, to respond to or ethically induce environmental guilt and shame. These conditions include relationality; trust in, credibility of, and authority of shamers; moral support; values; a vision of identity that takes account of human limits and failure; participatory processes; identifying with the other; critical self-examination and transformation; and action and ritual. The conditions are often dependent on each other and mutually reinforcing. The list is not exhaustive. Because maintaining all of these conditions at once is difficult at best, ethically intentionally inducing shame is *even more* challenging than suggested previously. Because these conditions are also needed to respond to guilt and shame, people should foster them regardless of their intent to induce guilt or shame.

Throughout the chapter, I emphasize the position of the potential shamer in assessing whether conditions in themselves, the broader community, and especially the shamee are sufficient to justify inducing guilt or shame. The agents involved in inducing or responding to guilt or shame may be individuals, groups, or collectives. I take this general position to outline the conditions under which intentionally inducing guilt or shame is potentially justifiable. Additional conditions may be required in particular cases.

Environmental Guilt and Shame: Signals of Individual and Collective Responsibility and the Need for Ritual Responses. Sarah E. Fredericks, Oxford University Press (2021). © Sarah E. Fredericks. DOI: 10.1093/oso/9780198842699.003.0008

8.1 Conditions to Deal with or Induce Guilt or Shame

Remember that while the self-conscious social moral emotions of guilt and shame are about judgments one has of one's actions or self, these emotions occur in a relational context: Another person or group (maybe including our self) is harmed because of our actions (Turner 1982, 15–18). Agents may also feel these emotions because of actual, possible, or imagined judgments about their actions or selves by others. Given the relational nature of these emotions, inducing or responding to them requires relationality. Relational responses may not need to be explicit, but in our generally individualistic culture, it is critical to pay attention to relationships (between any various combinations of individuals, groups, or collectives) when responding to guilt and shame. The following conditions identify characteristics of such relationships.

Trust between the shamer and shamee and credibility and authority of the shamer are also critical to successful induction of or response to guilt or shame (Jacquet 2015a, 105–12). The shamee is more likely to be shamed by one they trust and/or one who is credible and has moral and/or epistemic and moral authority. Without such features, reputational judgment (by one's self and others) and shared standards have little claim on the potential shamee. Responding to or inducing these moral emotions can also establish, sustain (and erode) trust, credibility, and respect between the parties.

Often such trust, credibility, and authority has been built over time in relationships—between family members, friends, neighbors, coworkers, members of a religious community or social organization. Thus, for example, Bingham's SUV driving parishioner was probably affected by her sermon about gas guzzling SUVs because of his previous relationship with Bingham as his religious leader, including her moral authority and credibility as well as the trust he had in her. Such characteristics may also affect the shamer. Bingham's commitment to her congregants led her to reconsider shaming. In relationships where the dignity of others is regularly honored, intentional shaming will stand out as unusual—either a signal that the shamer has overstepped or that the shamer is gravely concerned about something. Trust, credibility, and authority also describe relations between and among individuals, groups, and collectives.

Much environmental writing aiming to motivate behavioral change explicitly or implicitly fosters trust, credibility, and authority. Popular environmental authors often work to establish themselves as normal whether through comparisons with statistics about the average person in their nation; anecdotes about their neighbors and friends; or confessions of their prior lack of knowledge, concern, or environmentally-friendly behavior (Beavan 2009; Harding 2009; Hickman 2005; Pearce 2008, 4–6; Pleasants 2009; von Jessen 2009). Similarly, rhetorical strategies—whether storytelling, appeals to emotions, or the marshaling of scientific evidence—appeal to readers' notions of sound evidence and authority. For instance, in *Confessions of An Eco-Sinner*, Fred Pearce travels around the world to uncover the environmental and social impacts of his everyday actions such as buying fast fashion or eating the products of industrial agriculture. He establishes himself as one who has done the research and who has firsthand knowledge and emotional experiences of those with whom he engages. Thus, he positions himself as is an authority on social and environmental injustices (Pearce 2008). Having established credibility and trust with their readers, authors can name activities shameful and encourage their readers to see their actions, selves, and communities in this way. Authors also rely on these relational characteristics as they articulate ways to deal with guilt and shame, whether by avoiding behaviors that may precipitate guilt or shame, mitigating the effects of such actions, or dealing with the emotional and existential consequences of guilt and shame. Establishing such a rapport between author and reader (or any shamer and shamee) is particularly important to avoid alienating people with a holier-than-thou attitude, a common reason people reject shaming or even environmental judgments.

Environmental authors writing to everyday environmentalists often identify their own wrongdoing, shame, and struggle to live up to their own environmental standards even as they have reached some ways to limit such problems, a helpful corrective to Jacquet's argument that the people inducing shame need to be above reproach for the shaming to work. While moral authority is important, perfection is not required. Expectations of perfection can exacerbate totalizing shame rather than more nuanced approaches to human identity and capacities necessary to acknowledge our humanity and deal with guilt and shame. Of course, a significant discrepancy between actions and ideals can certainly undermine shaming efforts (Jacquet 2015a, 111), as seen in public backlash of Al Gore's environmentalism because he lives in a large air-conditioned home and frequently took planes to his climate lectures. Yet, expecting shamers to be perfect ignores the realities of how relationships of moral accountability often work.

Relationships characterized by trust, credibility, and authority also enable agents to serve as moral supports for each other. Such support can include the solidarity of shamers admitting to their own shame or the listening ear or cheerleading enabled by relationships between family, friends, coworkers, or members of a religious community or club. While support systems have existed for a long time, their intentional use has blossomed in recent years as people recognize how helpful they can be and as traditional support systems have eroded as people move and change jobs more frequently, as family size decreases, and as participation in social, civic, and religious groups decreases. Think, for instance, of the many popular social support programs for weight loss, increasing fitness, and dealing with addiction as well as emerging relationships of everyday environmentalists in online communities discussed in Chapter 2. Such support is also evident in competitive challenges in physical or virtual environmental organizations to cut one's carbon footprint, reduce food waste, or limit consumerism. These groups not only set goals and monitor progress toward them but also provide moral exemplars, cautionary tales, and education about how best to reach such goals, and emotional and existential support. Ideally, this process not only includes making goals and plans consonant with one's values but also enables agents to deal with their inevitable mistakes or partial successes as well as the need to reassess one's values and/or goals over time. In online environmental confessional rituals, respondents remind the ashamed that their shameful acts or existence is not the sum total of who they are, that they also do environmentally praiseworthy things, and that people in general are not in fact perfect. Underlying such reassurances is a vision of human nature and individual identity that has room for human potential and for acknowledging human limits and even failures. These ways of receiving support can raise agents' consciousness, help them imagine alternatives, and aid them in dealing with the psychological, existential, and social challenges of guilt and shame, even though the agents are limited.

We see a conspicuous lack of moral support in many cases of internet shaming where a person is singled out for wrong-doing and intentionally, viciously shamed (Ronson 2015).[1] The absence of such support is a strong indicator of why shaming by internet mobs is often problematic. These incidents rarely give any real opportunity for the shamed to apologize, change behavior, or reintegrate into the community, let alone support in doing so.

Moral support is also salient for collectives. Communal emotions require communal responses. Some environmental initiatives such as the Green Congregations Program recognize this need for when they introduce congregations interested in greening their physical buildings and worship to congregations who have already gone through the process. The more experienced partner provides concrete tips about implementing new physical and procedural initiatives as well as social support to stick with the initiative despite inevitable challenges. While this program is not focused on guilt or shame, it illustrates the power of moral support in collective actions as participants attempt to imagine and enact alternative ways of being.

Moral support is most commonly framed as critical for the agent who has a problem or is working to make a difficult change (the shamee), but it is also critical for the shamer. As Nussbaum points out so well, one danger of shaming is that the shamer will go too far, shaming the shamee beyond redemption or recovery

[1] These cases, well chronicled by Jon Ronson in *So You've Been Publicly Shamed*, are substantially different from the subtle induction of shame by blog authors discussed in Chs 2 and 7. Ronson's cases target specific people for intense shaming, while the cases I examine are not targeted at specific people and are not as relentless (Ronson 2015).

and/or ever expanding the scope of shame (Nussbaum 2004, 233–8). Additionally, it can be tempting for shamers to assume a position of superiority, failing to recognize their own limitations, as they increasingly nag, scold, and hold others to unreasonable expectations. The shamer needs methods of reigning in these dangers, often through relationships with other agents willing and able to question the shamer's harsh tactics, praise what is appropriate, talk through the grey areas, or model acceptable shaming. Explicitly considering whether the means of shaming align with values of shamed and shamer and provide true opportunities of support for the shamed can be helpful as can providing checks against the expansion of power of the shamer by promoting participatory processes of inducing and responding to guilt and shame (see below). Moral support for shamers may also be vital in situations of unintentionally induced shame so they can respond to the negative experiences that they catalyzed in others.

Moral support for these various agents associated with shame can happen spontaneously or be formalized. Given its importance and the risks of shaming, intentional moral support for shamer and shamee should be an explicit and necessary feature of any moral community, particularly those intentionally employing guilt and/or shame.

Inducing guilt and/or shame and responding to these moral emotions presupposes values. In general, for an agent to induce guilt or shame in another, the values of both shamee and shamer are salient. The shaming will depend on the shamer deploying values that the shamee holds to shame them. Similarly, responding to guilt and/or shame requires appeals to values in order to assess, contextualize, and act after these moral emotions. Frequently, values are shared because the shamee and shamer are in some sort of community—a nation, city, religious group, workplace or industry, club, neighborhood, support group, etc.—and have assented to, developed, and/or implicitly adopted such values over time.

While it is conceivable that the shamer could induce guilt or shame in the shamee according to values that the shamee, but not the shamer, held, this possibility sets the stage for manipulation. For example, marketers might prey on parents' desire to protect their children from harm as they shame them into buying "natural" products merely to enhance their clients' bottom line. Such manipulation is ethically problematic because of the unequal power differential and access to information between marketers and consumers in the contemporary world, as discussed in Chapter 7.

Recognizing that dissimilar values can facilitate manipulation, one may be tempted to require that the shamer and shamee fully share values to justifiably induce guilt and shame, as such an ideal situation would close off an avenue for manipulation (to change values). Yet, assuming completely shared values is too high of a standard. Even in the most homogenous relationships there will be differences in the interpretation and prioritization of values between agents (and often within a single agent). Requiring complete agreement will likely defer the possibility of shaming or inducing guilt (or any application of values) to an ideal future state, forcing participants into a never-ending discussion of clarifying values, rather than a pragmatic approach of trying to address problems through consensus based action (Norton 2005, 92–129).

Demanding complete agreement of values before action is particularly problematic in novel or developing ethical situations such as climate change which may require new ethics or the rethinking and extension of existing ethics. After all, people look to guilt and shame as tools in part because they can catalyze the development of moral resources. All values need not be shared, but agents supporting the shamed should at least be able to care for the shamed and recognize their values in order to promote their dignity and avoid problematic manipulation. Trust, credibility, moral authority and moral support will facilitate (and be facilitated by) the maintenance and development of values.

While a wide range of values could be touchstones for people responding to or inducing guilt and/or shame, without a vision of identity, it is hard to imagine that inducing shame could be justifiable or that one could deal with guilt or shame. One needs a sense of identity that allows for the possibility of wrongdoing, failure, and coming up short as well as the possibility of doing well, learning, and positive change. One's vision of identity and personhood should allow emotion, action, and being without being overconfident or wallowing in guilt or shame. Such a metaphysical anthropology could be rooted in God's grace or mercy, the divine spark in people, the lack of ultimate separation between entities, or one of many other ideas. Regardless of its ground, one's vision of identity must be able to make sense of failings and the possibility of success in the past, present, and future. Admittedly, such a concept may not be explicitly articulated as it may be a presupposition so deep that the agents do not think to name it.

It may also be helpful for a vision of identity to include a sense of how an individual should prioritize themselves vis a vis others with whom they are in relationship. Research on cultures which prioritize relationships over individual preservation indicate that this prioritization may enable people to avoid the paralysis or hiding of shame in order to take reparative action (Bagozzi, Verbeke, and Gavino Jr. 2003; Goetz and Keltner 2007).

Agents experiencing shame need these features of identity or they will not be able to understand whether and how the negative consequences of their actions can be checked or overcome. There will be no way out of the shame, no way to change their identity, no way to change their future interactions with those they have hurt, no way to hope they could do better. In other words, there will not be a way to experience guilt or shame at a part of one's self or collective rather than entirely whole-self-judging shame. Agents need the former view to be able to have and believe in a chance of change.

The shamer also needs such a vision of identity if there is to be anything short of complete ostracism of the one shamed. If the shamer does not think a marred agent has any possibility of change, there will be no potential to maintain the community. A view of identity that allows for both failure and success is also needed for the shamer's own sake. A commitment to such a view can help ensure that the shamer does not set unreasonable expectations for their own agency or capacities, ignoring their own fallibility. It establishes that their ideas and action plans will be impartial and imperfect. This may suggest that they should be open to correction and, ideally, able to be reversed or adjusted over time (Norton 2005, 92–129). Thus, while a vision of identity is needed to respond productively to guilt or shame, it is also critical to inducing these moral emotions insofar as such a vision provides a ground for avoiding extreme shaming.

In this discussion of the characteristics of relationships and circumstances necessary to responsibily induce or respond to individual, membership, and collective guilt and shame it may be tempting to overemphasize the promise of communities. Indeed, harmonious cooperation (or steps toward it) is frequently framed as an ideal vision in environmental ethics.[2] Relationships in communities are critical to ethically inducing and dealing with guilt and shame, but we must recognize their limits. Communities often induce guilt and shame based on inconsistent norms and/or judgments or actions that belie the dignity of those involved in ways that may have devastating psychological, social, and physical impacts on individuals experiencing guilt and shame, those inducing these experiences, and the society at large (Jaspers 2001; Jordan 2003, 44–6; Nussbaum 2004). Thus, we also need a process of responding to and inducing guilt and shame (if they are to be ethically justified) that can counter such extremes. Participatory processes are one option.

Participatory processes are those in which all affected by actions have the opportunity to participate in the meaningful formulation of, decision-making about, and enacting of the actions. In one sense, participatory processes always pervade initiatives to induce or judge guilt and shame because unless people agree on some values and/or accept the moral authority of the other, they will not feel guilt or shame. Yet, such participation is often insufficient, pseudo-participation because the powerful people dominate others. To avoid, or at least minimize, coercive power plays and honor the value of all participants, a participatory process open to revision, in which the person or people inducing the shame are not outsiders, or overlords, but of the community is needed. In this process, no one is above critique, but all are subject to it. The process is one of invitation and persuasion rather than coercion. The power dynamics should be such that the agents inducing and experiencing guilt and shame are able to

[2] See Lisa H. Sideris' *Environmental Ethics, Ecological Theology, and Natural Selection* for documentation and critique of this trend (Sideris 2003, 24–5, 45–90).

consider, reflect, and criticize the values and process of shaming individuals or collectives.

These ideal participatory processes are different from cases of marketing using guilt and shame in the previous chapter and Nussbaum's analysis of state shaming of individuals. In those cases, shame or guilt is induced to change the behavior of the relatively less powerful to reach ends determined by the powerful without consultation or consideration of the needs and values of the shamed. Participation is preferable when it builds on and builds up shared values; considers how values extend to new situations; enhances understanding the outcomes of actions; and enables agents to recognize their limits without being devastated completely by them. In these cases, people can help identify new values, critique old ones, and decide which to live out. We've seen examples of how shame can prompt reflection on one's own identity and values. Participatory responses to guilt and shame are not just an erasure of the wrongdoing or problematic identity but a rich, nuanced, and transformative response. Participation enables the transformation of agents who are guilty and/or ashamed and those who identify with them. These participatory actions may happen consciously or unconsciously, but given the power dynamics in our world, rich participation is unlikely without intentionally fostering it.

While one might wish for a universal, clear line between genuine and acceptable participation and between sufficient and insufficient participation, such a line cannot be drawn without running the risk of being domineering rather than enabling participants to have a say in the process itself based on their priorities. The line, such as it is, must be established through participation given the participants' vision of the collective, the historical and contemporary circumstances, and ideal of community and or collective support. Furthermore, this line will need to be open to revision over time as people have new experiences, get new data, as they are convinced of new priorities, reevaluate existing values, and as participations enter or leave the process. Different individuals and groups may accept different levels of force with which guilt or shame is induced depending on the level of trust between participants, the degree to which they intentionally or explicitly look to the other for accountability, the moral experience and authority of the shamers, and the level of and specifics of shared values between them. When and if guilt or shame are induced, significant care must be taken, even if the process is participatory, built on trust, and out of shared values, to ensure that the most negative consequences of guilt and especially shame are avoided or, at minimum, are able to be dealt with well.

One potential adverse consequence of shame is the narcissism that it can promote. Shame does not, however, have to stop an agent's gaze at their own navel. Alternative outcomes arise as shaming emotionally and cognitively draws attention to a previously underexplored gap between a) an agent's character and b) ideals. When this recognition is coupled with emotional and rational identification with the one harmed, the shaming can direct one to look beyond oneself to others, "reinforcing a sense of common human vulnerability" (Nussbaum 2004, 213; 2013, 187).

Nussbaum makes this point when considering her rare example of the possibility of constructive shaming, centered on her analysis of Barbara Ehrenreich's book *Nickle and Dimed: on (not) getting by in America.* Ehrenreich describes how, in order to learn about the experience of the working poor she tried to support herself through minimum wage jobs without relying on her education, advanced skills, or personal networks. Meeting her basic needs was difficult under these circumstances. Coupling these experiences with broader research led Ehrenreich to argue that the plight of the poor working-class people in America was not due to laziness or personal faults but rather due to systematic infrastructural, societal, and ideological conditions. She maintained that the United States' lack of support for the working poor is a moral failure (Nussbaum 2004, 211). When considering the appropriate response to this state of affairs she wrote "[G]uilt doesn't go anywhere near far enough; the appropriate emotion is shame" (Ehrenreich 2001, 220–21).[3] Nussbaum argues that this quote demonstrates Ehrenreich call for U.S. Americans to "reexamine our habits and our national character" not just "apologize for this or that harmful action" (Nussbaum 2004, 212).

Shame is operative in Ehrenreich's challenges to U.S. Americans to change our habits and character, Nussbaum argues, because Ehrenreich draws attention to the vulnerability shared by all humans, (e.g. the vulnerability of limited physical energy as well as the need to eat, sleep, have shelter) and the limits of existing in a physical and social system that hinders fulfillment of these needs, making many people's life miserable (Nussbaum 2004, 213). Ehrenreich invites and enables readers to empathize with the working poor and reflect upon how they would fare in similar circumstances. Such identification could spur self-absorption if one became obsessed with how one participates in circumstances that directly or indirectly harm the other or reflect one's character. However, the coupling of such identification, empathy, and judgment of wrong can also have at least two positive outcomes. First, it can broaden the sphere of concern beyond oneself to include the other and so can counter narcissism that can accompany shame.

Second, it can provoke "critical self-examination" in which one reflects on how one contributes to exacerbating the other's vulnerability and whether this is the kind of person one wishes to be. From this "critical self-examination," reform is possible or likely for Nussbaum, because "Americans should examine their ways of life and their commitments, and so doing, realize with shame that we have

[3] While Ehrenreich's book called significant attention to the working poor and serves as a good model of how an author can shame readers for being a part of the system that contributes to these conditions, she sometimes also looks down upon the working poor. Thus, her work is not yet an ideal model of recognizing and supporting the dignity of all people.

failed to live up to ideals of equal respect that are central in our society" (Nussbaum 2004, 241; 2013, 364).[4] Such action can have a variety of positive implications including catalyzing political activism and avoiding the hiding and denial associated with shame.

Critical self-examination arising out of shame begins with experience that one (an individual or collective) is not living up to one's standards. While this is often discussed as a simple failing to meet a threshold to reach acceptability or crossing a line into the unacceptable, many other variants can exist. One can realize (explicitly or implicitly) that one's values have a broader scope of applicability than one thought or that what one values is not served by one's actions. One may also experience that one holds multiple values which appear to conflict when applied to the case at hand, but that in recognizing the potential conflict and acting in spite of it one comes to understand the kinds of compromises one is willing to make and therefore which values one prioritizes, at least in that case. Alternatively, one may find that one's values are still commendable but one's will is weak or that the physical and social system in which one is enmeshed hinders one's possibilities. Especially in these more complicated situations, one needs to be able to implicitly or explicitly assess whether one can live up to one's values, an analysis that depends on one's theories of identity and human nature that is shaped by and shapes how one lives one's life including one's thought, emotions, relationships and physical acts. All of this is not a solitary experience but is communally shaped. Moral support, a strong self-identity embedded in a vision of humanity, and other features discussed above are also critical to this process.

Shamers should critically examine their identity and actions to ensure that they are not promoting themselves and their priorities to the detriment of others and that they not only induce guilt or shame in others but also follow through with the support the other needs. This ongoing reflection should draw upon feedback and moral support from the community, rooted in its values and a vision of identity that encompasses capabilities and limits. It will be aided by a recognition of the dignity of the other, and possibly, identification with them. Doing all of this will enable self-restraint in the process of shaming. Sometimes self-reflection will lead to a decision to stop shaming as in Bingham's account, but it does not need to.

Critical self-examination and transformation should not be a singular event. Rather, like the larger process of eliciting and responding to shame, it is an ongoing activity. Practice enables one to develop and maintain some degree of proficiency. While these practices should involve reflection to ensure they align with one's values and sense of identity and have a reasonable chance of being effective,

[4] This shame, arising from a mixture of actions and identity, is a step away from the pure self-oriented shame based on finitude or abnormality that Nussbaum usually focuses upon. In other words, Nussbaum shifts to a different form of shame right when she says shaming may be valuable. This move opens up lines of analysis similar to what I have been discussing.

they are not limited to conceptual processes of reasoning or evaluation. They should ideally infuse, guide, and be reinforced by how the agent (shamer or shamee) acts. Acting is necessary not only as an outcome of guilt but to move from shame towards guilt because as we act we change who we are.

Thus, responses to guilt and/or shame may also include actions aiming to mitigate the negative consequences of past actions—replanting trees or picking up trash. Response may also set up conditions so it is more difficult to repeat past deviations from the ideal, i.e. by enacting and enforcing environmental laws. One may also investigate where to buy organic foods, sign up for action alerts about proposed legislation, or choose to live close to work. A community or nation could raise fuel efficiency standards or invest in bike lanes, sidewalks, and public transportation. Responding to guilt and shame can also involve apologizing to those harmed or making financial, relational, political, or environmental restitution (even though such acts are only a partial response). It can involve symbolic enactments of alternative futures through fiction, theater, song, dance, or other forms of artwork. None of these activities are mutually exclusive and it is likely that all of these types of actions will be needed at various points.

These activities involve our bodies and physical circumstances as well as the emotions and rational reflections intertwined with them. We need this movement, this action, often in a sustained or repetitive fashion, to influence our identity, to become who we are and will be as individuals and collectives. Ideally, such transformation is not just of the person, membership group, or collective agent who did or was wrong and was shamed, but of the entire community including the shamer. Relational responses are necessary when the problems are also relational.

Such transformative action can help limit particular problematic actions and the conditions that lead to, facilitate, and normalize such behavior. In the next chapter, I draw upon the work of Jordan and his colleagues, other ritual theorists, and examples of environmental rituals to argue that rituals are forms of action that are particularly potent at meaning-making and identity formation. Rituals can also foster the conditions for dealing with and inducing guilt and shame. But first, I should consider how the set of conditions influences my answer to the question of whether intentionally inducing guilt or shame can be ethically justified.

8.2 Can We Justify the Induction of Guilt or Shame?

The list of conditions for dealing with and inducing guilt and shame articulated in the previous section is lengthy. Satisfying all of them simultaneously is highly unlikely. Thus, anyone convinced that these conditions are necessary has one more reason to be cautious about intentionally inducing guilt or shame. Yet, I hesitate to conclude that inducing guilt or shame can never be justified. Requiring perfection in order to use any tool inevitably means, if one is honest, that no tool will ever be used because humans are limited beings. A standard of perfection is therefore out of line with reality. It also reinforces unrealistic ideas of human nature that exacerbate the experience of guilt and particularly shame and make it difficult if not impossible to deal with them productively. Rather than expecting perfection from the outset we need to be able to live into the ideals developed as we act, react, feel, and evaluate as individuals and collectives in relationships. When we take actions we know are in tension with some of our values we should have (a) regret that things have gotten so bad to warrant such risks even as we recognize that we will always be limited in implementing our values, (b) a commitment to creatively exploring other strategies to achieve similar ends, and (c) a willingness to adapt our approach to lessen its negative effects as the situation changes or as one's knowledge increases. In other words, I advocate a humble, cautious, adaptive strategy of employing guilt and shame if they are to be used, though I do not recommend that they are used often.

New complex collective action problems such as climate change may demand the development of values, practices, and identities (of individuals, memberships, and collectives) that are as of yet insufficiently conceptualized let alone instantiated (Jenkins 2013, 2014). While guilt and shame are far from the only way to accomplish this, using these tools along with pride and praise for a job well done (Jacquet and Dyson 2014, 110, 164–5) and coupled with best efforts to develop and live out the participatory conditions indicated above, is one possibility. These are tools humans instinctively use. To reject them because of real but not insurmountable ethical concerns is to hamstring us in a crisis rather than keeping all of our options open.

Furthermore, given the pervasive nature of guilt and shame in human experience and their role in the development, maintenance, and living out of values and identity and the way these experiences can occur without intentional inducement, it is unlikely that guilt and shame experiences could be eliminated. Whether or not people intentionally induce guilt or shame, some will feel these moral emotions. Since we need the same conditions to deal with guilt and shame as to induce them, we should work to foster such conditions even if we do not intend to induce guilt or shame. Thus, we come to a final question: how can individuals and collectives bring about and sustain such conditions? In the next, final chapter, demonstrate why rituals are particularly well suited to such tasks.

9

Ritual Responses to Environmental Guilt and Shame

Having articulated the conditions to respond to or induce environmental guilt and shame, it is reasonable to wonder how humans develop such resources. We have already seen that the everyday environmentalists are beginning to do so. Everyday environmentalists gather and constitute communities on and offline, request aid in dealing with guilt and shame, provide moral support to their peers, and suggest modes of action to transform their identities and respond to the consequences of their actions. They subconsciously create community and religious rituals to deal with their environmental guilt and shame (Fredericks 2014b). Their turn to ritual suggests that religious studies scholarship, especially of the intersection of ritual and ethics, will be a particularly useful mode of analysis to answer this question. Indeed, William R. Jordan III writes

> At the deepest level, ritual offers the only means we have of transcending, criticizing, or revising a morality or ethical formulation prescribed by authority or handed down by tradition. Most fundamentally, it is the means by which humans generate, recreate, and renew transcendent values such as community, meaning, beauty, love, and the sacred, on which both ethics and morality depend. In the absence of both a well grounded ritual sensibility and a repertory of rituals adequate to the various tasks at hand, the lower values of economics or power or top-down appeals to ethics and other prescriptive systems erode and override these higher values in the ongoing struggle to define what is valuable and how people should behave. (Jordan 2003)

If Jordan is even partly correct, ritual is a promising resource for developing and maintaining the conditions articulated in Chapter 8. I will demonstrate in this chapter that rituals not only have the ability to foster values, identity, community, and action as Jordan suggests but also to create and sustain the other conditions.[1]

To ground this argument, section 9.1 explores the basis of ritual theory, relying on the work of Jonathan Z. Smith and Catherine Bell, particularly regarding rites of affliction, which respond to disorder or wrong. Section 9.2 turns to

[1] This chapter will focus upon responding to guilt and shame given their perennial existence and due to the ethical challenges of inducing guilt and shame.

Environmental Guilt and Shame: Signals of Individual and Collective Responsibility and the Need for Ritual Responses. Sarah E. Fredericks, Oxford University Press (2021). © Sarah E. Fredericks. DOI: 10.1093/oso/9780198842699.003.0009

environmental ritual, including the theories of Jordan, Todd LeVasseur, Gretel van Wieren, and Joanna Macy, as they see rituals as ways people respond to negative experiences, affects, and states of being. While their work bolsters my argument that rituals are particularly relevant ways to respond to environmental guilt and shame, their arguments do need expansion to deal with relationships between humans or involving collectives. Section 9.3 demonstrates how spontaneous environmental confessional rituals by everyday environmentalists in popular online media and an apology ritual at the Standing Rock prayer camp can attend to these relational and collective issues. Yet, as I recognize that these rituals do not fully embody the conditions and thus do not respond to all of the dimensions of guilt and shame, I will briefly explore the need for and limits of intentional ritualization in the final section.

But first, a brief interlude about the term "religion." Religion encompasses traditions such as Judaism, Christianity, Islam, Hinduism, Buddhism and Indigenous traditions, as well as new religious movements, civil religions, and nature religions, which share a number of characteristics such as doctrines, rituals, narratives, emotions, ethics, social structures, and material objects (Smart 1996) and consist of symbol systems that are taken as real by their practitioners (Geertz 1973). Defining religion broadly in these ways allows us to consider aspects of human life such as that of online confessions of everyday environmentalists which can be illuminated using modes of analyses from religious studies even though the participants may disavow religion or fail to identify their activities as religious behavior (Fredericks 2014b).[2] Casting our net wide is particularly important in this project because we cannot presume that a response from one religious tradition will be sufficient for all and because traditional religions do not adequately acknowledge let alone deal with environmental guilt or shame. Thus, I aim to articulate a general form of ways to respond to guilt and shame, not mandate the details.

9.1 Ritual Theory

When using the term "ritual" I similarly employ a broader understanding than is common in the vernacular (Fredericks 2016). Bell, a leading ritual theorist, notes that modern Western people typically conceive of rituals as formal, traditional, governed by rules, and static. They generally presume rituals involve sacred symbols and connect practitioners to the divine or transcendent (Bell 1997, 138). Yet, Bell rightly argues that rituals are not necessarily limited to these characterizations. She maintains that rituals actually include a much broader array of actions

[2] In their 2018–19 telephone study, the Pew Forum found that 26 percent of adult U.S. Americans were unaffiliated with a religion (Pew Research Center 2019).

and that ritual is at heart "a way of acting that differentiates some acts from others" rather than any particular set of special practices (Bell 2009a, xv, 74). Such differentiation may occur between people, places, roles, emotional states, moral judgements, etc. Likewise, J. Z. Smith counters modern Western, especially Protestant stereotypes about ritual, when he asserts that ritual is "a mode of paying attention . . . a process for marking interest," not of blind action or habit but of attentiveness (Smith 1987, 103).

During rituals participants may pay attention to a new stage of life, events they wish to commemorate, or processes of healing or teaching. While the phrase "paying attention" may suggest cognitive processing, rituals are not limited to thinking and may not prioritize it. They can also involve producing and experiencing sights, smells, sounds, and tactile experiences as well as a wide variety of movements or the lack thereof. Participants, whether individuals or collectives, often feel significant emotions, including awe, wonder, fear, joy, ecstasy, grief, shame, pride, etc. in ritual. They are often places of emotional contagion, as when people experience the collective emotion of a wedding or funeral. In short, rituals may involve all dimensions of human experience: cognition, sensation, and emotion.[3]

Rituals frequently facilitate or accompany change. They can spark new experiences of familiar things, including the body, emotions, and relationships to others or enable new experiences i.e. of a new life stage, role, or place. They may also increase awareness of or connection to the transcendent. Rituals, and the transformation they engender, may be repeated often such as the reconciliation after an apology or forgiveness, or may be singular experiences such as conversion or baptism.

Regardless of the type, all rituals functionally maintain and create identity for the individual and communities involved in them. Identity is often developed in contrast, for instance, with one's own pre-ritual self, or with people who do not participate in the ritual. Rituals may characterize participants as those who do, believe, or know something; are in or from a place; or who have a particular responsibility, ability, or limitation. Identity formation can touch the very core of the existential sense of self, of what it means to be oneself in particular, a part of a particular community, or a human in general. Insofar as rituals build the identity of a collective and foster relationships between agents, they can cultivate trust, credibility, and authority, as well as provide conditions of moral support between people. Furthermore, because ritual participants are frequently supported in their practice by their community and often experience this identity formation as a group, their identity is bound up with that of their fellow participants. Rituals may also form collectives whose identity is more than the

[3] I name these experiences separately to ensure each is included, not to presume sharp divisions between them.

sum of their parts. All of these types of identity formation can respond to existential questions raised by guilt and especially shame. In other words, through ritual, agents can change their behavior, and/or being as is required to respond to guilt and shame.

Differentiation, the key aspect of ritual for J.Z. Smith (Smith 1987, 109), not only occurs in identity formation but also when ritual differentiates the ordinary and the ideal. Whether dancing, acting out stories, singing, or performing other acts, rituals identify what is ideal and enact it in some way (Smith 1987, 109). In this process, people at least implicitly acknowledge their limits, whether a particular problematic action, limited relationship with others, or even the existential limitations of self or species. Rituals frequently also provide means to bridge the gap between the actual and ideal. Bell names rituals which "attempt to rectify a state of affairs that has been disturbed or disordered," rites of affliction (Bell 2009b, 115). Jordan recognizes the importance of such rituals "as the means by which a society manages discordant and divisive elements, both within the individual and among individuals, in order to achieve a measure of harmony, experienced as a value such as community" (Jordan 2003, 35).

Many religious traditions employ rituals to deal with limitations they presume humans have. Frequently, such rituals acknowledge the wrong and may punish offenders but maintain that the wrongdoers are still members of the community and have existential worth. In these cases, rituals envision and maintain community and individual identity. They chart a path in which the harm should be lessened (e.g. the day of Atonement in Judaism, or processes of relinquishing or letting go of perceived transgressions in Hinduism or Buddhism). Barend J. de Klerk argues that traditional Christian liturgies recognize the need for lament, grief, and confession generated by humans continually failing to live up to their own ideals. Yet, de Klerk notes that liturgies also express praise and hope as they continually renew the possibility of something better (de Klerk 2014). Thus, liturgies ideally contend with the negative emotions associated with failure, loss, guilt, and shame without becoming stuck in them. Through these rituals, participants learn, develop, and embody, not just think about, existential explanations which are woven into the sense of self, community, and, potentially, species identity through the ritual. Thus, rites of affliction can directly address the existential, identity related issues arising with guilt and shame.

When rites of affliction involve symbolic acts that can inspire future concrete acts and/or concrete actions in and beyond the ritual space they may repair wrongs and prevent future harms. For instance, rituals may include apologies to those harmed by environmental degradation, planting trees, installing energy saving appliances, and advocating for environmental legislation or blessing bikes which may indirectly inspire and validate their use (Everett 2017, 163–9). Thus, rituals enable the transformation of identity, establishment and sustaining of relationships, and the transformations of action needed to deal with guilt and shame.

While the above discussions of ritual focus upon how they identify and develop individuals' identity and shape their actions, all of these things can and do also happen in rituals for collectives. Rituals are particularly appropriate ways to shape identities and actions in collectives insofar as they often involve collaborative events and, as such, can constitute and reinforce a collective. Even if rituals are explicitly focused on an individual (e.g. a coming of age ritual) they can still reinforce and bind together the community who believes and practices related ideas and delineate the individual with respect to the community.

Admittedly, just because rituals can foster the conditions necessary to respond to environmental guilt and shame does not necessitate that any particular ritual, especially as enacted in the world, will meet these conditions. Rituals can stultify with little to no opportunity for critique or challenge by participants. Rituals may be forced on participants to promote particular ideas, identities, or action through various coercive or manipulative means. Cultural suspicion of ritual, influenced by enlightenment and Protestant ideas, may also hinder ritual from aiding responses to guilt and shame. When we consider rituals accepted by their communities and practiced in ethical ways, we must recognize that they may not meet our needs. Many rituals are not rites of affliction, and of those that are, many do not address environmental issues let alone environmental guilt and shame. Despite these possible, even frequent, limits of rituals, they can still be meaningful and useful. Employing them with care and conscious effort to make sure that they are ethical, participatory, and effective may be the only way to address all of the dimensions of environmental guilt and shame.

9.2 Rituals and the Conditions Necessary for Inducing or Responding to Guilt and Shame

With this basic understanding of rituals, we can now explore how they can foster the conditions of inducing or responding to *environmental* guilt and shame by examining theories of environmental ritual. Jordan, LeVassuer, and their collaborators argue that environmental restoration is a form of ritual, which, among other functions, helps people deal with environmental shame. Their work illustrates the connection between rituals and conditions of values, identity, and action. Van Wieren and Macy's work indicates that ritual can spark self-transformation. Yet these theories as a group require elaboration to illuminate the connections between ritual and interpersonal relationships as well as applying the conditions to collectives to large spatial temporal and temporal scales (Jordan 1993; 2003; Jordan, Barrett, and Curtis 2012; LeVasseur 2011; Macy and Brown 2014; Macy and Johnstone 2012; van Wieren 2013).

Jordan and his collaborators examine shame as a critical element of environmental restoration, "*everything we do to a landscape or an ecosystem in an ongoing attempt to compensate for novel or 'outside' influences on it in such a way that it can continue to behave or can resume behaving* as if *these were not present*" [emphasis in original] (Jordan 2003, 22). Jordan's prime example of restoration is prairie restoration wherein people plant seeds, pull invasive plants, and spark controlled burns to facilitate a return of prairie ecosystems in disrupted environments.

For Jordan, restoration is a form of ritual, "the means by which humans generate, recreate, and renew transcendent values such as community, meaning, beauty, love and the sacred, on which both ethics and morality depend" (Jordan 2003, 5, 148). Rituals enable societies to deal with change in life (49), the existential crisis of killing to eat or survive (45), and societal discord (35). Restoration creates meaning and value as people recognize, develop, and live out their relationships to the land (21–4), while also building morality (Jordan et al. 2012, 294–5; LeVasseur 2011, 31–3; van Wieren 2013, vviii–ix, 15–16). These dimensions of restoration rituals enable them to respond to shame (Jordan 2003, 50; LeVasseur 2011, 26–8).

Jordan conceives of shame as a basic existential identity (Jordan 2003, 46–7, 60; Jordan et al. 2012, 294–5, 299; LeVasseur 2011, 26). Guilt, in contrast, is

> the emotional register of a debt that might be repaid, or at least a failing for which we are responsible and of which we are 'in charge.' Shame, on the other hand, is the emotional register of our natural, radical, existential dependency and a debt for which we are not responsible and which we cannot repay.
>
> (Jordan 2003, 47)

The inability to pay for one's wrongs or dependency reveals the existential nature of shame. More basically, shame for Jordan encompasses the discrimination, differentiation, and/or categorization of things, since in all of these processes we are made "aware of limits" in our selves and our conceptualizations of the world (Jordan 2003, 140; LeVasseur 2011, 35). He is particularly interested in the shame of killing in restoration work, e.g. the killing involved in weeding or even nurturing some plants and animals over others though he and LeVasseur also recognize the shame involved in killing in general and in "experiencing [of] our inequitable economic dependency on nature" (Jordan 2003, 39, 42, 45, 88, 147; LeVasseur 2011, 28). Restoration rituals plainly and explicitly acknowledge these distinctions between what we value and what we do not. Restoration can make people confront their shameful participation in discrimination and killing (Jordan 2003, 185–8). Restorers learn and experience how individual entities, species, and ecosystems flourish (or fail to do so) in and outside of restoration processes (79); recognize the limits of their actions and power; learn that humans are not "superior to nature," they are a part of the system; and discover more intimate details about the ecosystem, including its dynamism (Jordan 2003, 120–23, 128–36), (LeVasseur 2011, 34). Thus, for Jordan, restoration is a way of recognizing and dealing with limits knowing what one can and cannot do (Jordan 2003, 140; Jordan et al. 2012, 301). In other words, restoration ritual responds not only to our problematic actions but to our existential crises. Its aim is not to "end killing or eliminate all wounds" but to "turn the act of killing" into "an occasion for communion" rather than "mere casual or unreflective killing in a clinical or matter-of-fact spirit" (Jordan 2003, 174). Clearly, Jordan and his colleagues see rituals, particularly ecological restoration as a rite of affliction that enables participation to develop shared values, take action, and develop a vision of identity.

Jordan and his collaborators also recognize the way that rituals can facilitate relationships. They, however, focus on human relationships to the land over relationships between humans (Jordan 2003, 61, 71–3, 88–90, 120; Jordan et al. 2012; LeVasseur 2011) to counter the extreme anthropocentrism in much of Western thought and action, which can exacerbate environmental degradation (Jordan 2003, 33–9; Jordan et al. 2012, 292–3). While this is important, relationships between humans are also valuable in their own right and because conditions of trust, credibility, and moral support are necessary to deal with guilt and shame.

These conditions are features of human relationships cultivated over time through repeated encounters. Preparing for and performing rituals can facilitate these conditions. One finds out who is trustworthy enough to gather the supplies, show up on time, take the proper attitude, etc. One learns who is supportive during the preparation for and performance of the rituals, especially arduous ones. Rituals also enable people to develop these traits as participants implicitly learn from role models, from explicit instruction, and practice. Arguably trust, credibility, and moral support only ever emerge over time. In communities with low levels of trust and moral support, specific rituals may be needed to facilitate these conditions. For instance, rituals of apology, restitution, punishment, and behavioral change may help establish trust between those who perpetrate anthropogenic environmental degradation and those who are harmed by it if such rituals enable perpetrators to recognize one's dignity, mitigate the negative effects of their past actions, and transform future actions to avoid harm (Conradie 2008; Fredericks 2016). Moving beyond small groups, families, a religious community, or a local group of ecological restorers, to larger spatial and temporal scales (i.e. about climate change) will likely need new relationship-building techniques to deal with the layers of individual and collective relationships involved, since most restoration practices, and ritual analyses of them, are focused on local, relatively small-scale projects (Jordan 2003, 110–13; Jordan et al. 2012, 310; LeVasseur 2011, 37–8; van Wieren 2013, 2). Such rituals may also enable perpetrators to deal with their shame and avoid shame spirals. Elsewhere, I have begun to explore what apology and forgiveness in a time of climate change may require, but much more theoretical and practical work on these subjects is needed (Fredericks 2019). It is, however, beyond the scope of this project.

Specific types of rituals are likewise necessary to foster the other conditions (participatory processes, identification with the other, and critical self-evaluation and transformation). Often intentionally fostering these conditions in ritual will be necessary to bring them about. For instance, van Wieren rightly maintains that restorative rituals should emphasize the full participation of community members, especially enabling political participation among groups that are often marginalized (van Wieren 2013, 116–17). She advocates continually asking about who is participating, invited to participate, and authorized to make these decisions (126), questions that could become a part of the ritual itself. Intentionally fostering participatory processes is likely necessary to counter the problematic power dynamics of much ritualization. It is particularly important when rituals seek to address shame which is often catalyzed by and maintains power differentials. Encouraging participation among the shamed is, however, likely to be difficult given shame's tendency to provoke denial, anger, a sense of low self-worth, and disengagement. Here, moral support, trust, credibility, and rich relationality among agents will be necessary to support intentional participation in ritualization.

Intentionality is also key in fostering identification with an other and avoiding the narcissism of shame. Van Wieren highlights the ways that the hearts of individuals, their very selves, can be transformed through participation in restorative rituals such as burning a restored prairie, planting seeds, or weeding (van Wieren 2013, 4). Through these actions they come to know the land and enrich their sense of self as one in connection to and dependent on the world and able to affect positive change in it. This transformational process enables participants to foster new values through ritual action (e.g. service to others) that can then be the basis of a wider environmental ethic (116, 126, 128–34). Such transformation is not merely an individual affair because restoration occurs in communities of people doing the work and possibly in the wider community. Yet, it is not clear that participation in ritual will necessarily or only prompt participants to recognize they can positively change the world. It is likely that, as Jordan notes, restoration work will also cause people to recognize their shame. Thus, intentional reflection on the transformative possibilities of restoration work is likely necessary to facilitate reflection on its meaning, build a vocabulary to discuss it, and enable ways to live it out in order to catalyze positive transformation including the ability to deal with guilt and shame.

Macy's environmental rituals, the Work that Reconnects, are a particularly strong resource for self-assessment and transformation rituals, though her work explicitly responds to negative emotions other than shame or guilt. Developed and implemented through a multi-decade long workshop series as well as in several books (Macy and Brown 2014, 68–9; Macy and Johnstone 2012), Macy's ritual framework aims to transform individuals through four steps: recognizing and/or expressing gratitude "for life on Earth"; recognizing and "honoring our pain," grief, despair, and possibly shame about environmental destruction; envisioning a new future; and finally putting these ideas and possibilities into practice. Because humans will inevitably act imperfectly, the four part cycle must repeat (Macy and Brown 2014; Macy and Johnstone 2012). Each of these four parts may be exemplified in a number of rituals. For instance, a ritual for the second step may include finding a natural object representing what one mourns, describing it to others, and then formally saying goodbye to the object and what it represents while others witness and support one's grief (Macy and Brown 2014, 118). Macy's rituals register the depth of negative feelings one may have regarding environmental devastation, whether caused by oneself or others. They aim to acknowledge and deal with such experiences without getting a person stuck there—to move to "active hope" as they foster communities of like-minded people to transform the world (Macy and Johnstone 2012).

Macy's rituals do, however, need extension along several key dimensions to be able to deal with the various challenges of environmental guilt and shame, particularly related to global issues such as climate change. First, they, like the others examined in this section, need supplementing to deal with the interpersonal relationships marred or broken by harms caused to another due to climate change or other environmental devastation. As described, the rituals do not include any direct apology to those harmed aside from oneself. If there is apology in the ritual, and from what I can tell, it is not a big piece, those harmed are not present to hear it. Yet, a direct exchange with those harmed may be important to build new community involving everyone, not just those that also feel bad about environmental destruction. Such interaction can help people recognize the inherent value of those harmed and the details of their experiences. It is also necessary to deal with the justified anger and resentment that people harmed by anthropogenic environmental destruction may feel toward those who have not acted to stop it despite their power and/or knowledge (Fredericks 2019, 145). Additionally, while Macy's work addresses individual transformation very well, she does not consider collective identity or action as will be needed to address collective guilt and shame. She emphasizes individual transformation of emotion, belief, and understanding so much that it is not clear how these changes will translate into concrete work to change the political, economic, social, and infrastructural systems outside of the ritual space as will be necessary to combat environmental challenges including climate change. She seems to think, alongside many scholars of religion and ecology, that the transformation of the self will naturally lead to these changes. While this could happen, systematic political, economic, and institutional factors can significantly limit what an individual or even a small group can accomplish. Additional rituals, or an extension of the ritual space, will likely be necessary to recognize and deal with the concrete temporal and spatial dimensions of climate change.

As agents intentionally develop rituals to deal with environmental guilt and shame, they should also attend to the ethics of doing so as discussed in Chapter 7. Effectiveness deserves explicit attention as it may be overlooked. It can be tempting when responding to guilt or shame to seek rituals that enable one to feel better quickly, by absolving one of wrongdoing, or reframing, even dissolving the problem. Yet, moving too quickly will likely fail to address their deep-seated emotional and existential concerns as well as the physical, infrastructural, political, and social circumstances which give rise to guilt and shame. For instance, if rituals quickly wipe away the bad feelings sparked by using plastic bags at the grocery store, agents will be left particularly vulnerable to shame whenever they realize these connections. Thus, a concern for effectiveness should prompt rituals to simultaneously consider emotional, existential, relational, physical, and other aspects of the situations giving rise to guilt and shame at various temporal, spatial, and agential scales. Concern for effectiveness will also likely push us to rituals operating at spatial scales larger than the local, especially when the issues that prompt guilt and shame are rooted in larger spatial and temporal scales. In sum, the complexities of environmental guilt and shame demand multiple, multifaceted ritual responses.

9.3 Environmental Rituals Responding to Guilt and Shame

Having examined how theories of *environmental* rituals developed in tandem with ecological restoration rituals and Macy's "Work that Reconnects" exhibit many conditions needed to respond to environmental guilt and shame while needing more development with respect to human relationships, collectives, large temporal and spatial scales and multifaceted effectiveness, we can now look to two contemporary environmental rituals which begin to fill some of these gaps. The existence of 1) online confessions of environmental guilt and shame and 2) military veterans apologizing to Native Americans at the Standing Rock Sioux Reservation prayer camp speak to the need their participants feel to have a way to deal with their environmental guilt and shame. While these rituals have some advantages over the rituals discussed in the previous section, particularly with respect to their attention to collectives and relationships between humans, they too need more development to fully meet the conditions.

Confessions of environmental guilt and shame, as described in Chapter 2, are spontaneously emerging rituals (Fredericks 2014b). They are often catalyzed by an everyday action such as using plastic bags at the grocery store, driving, or eating meat. This action, coupled with knowledge of ethical ideals and the action's environmental effects, makes the agent conclude they are and/or have done wrong. Then the agent confesses on a blog, in a comment on a blog, on a discussion board, in a book, or between friends in meatspace. After the confession, the confessor generally contextualizes their wrongdoing and another agent further contextualizes the guilt or shame in a larger theory of human identity and with respect to the agent's other past and potential actions. The guilt and shame expressed in the confession can be states, feelings, or some combination thereof. Responses from the confessants and their interlocutors in online dialogues acknowledge states and feelings of guilt as they propose reparative actions. They also respond to shame by articulating ways of conceiving of the self and human nature that allow for hope of doing and being good despite one's flaws and failures. They aim to right emotional, existential, and practical wrongs. While the particular activities and identities discussed are often explicitly that of the individual confessant in online conversation, they do imply collective contexts. Furthermore, memberships and collectives are more explicitly acknowledged in longer or deeper forms of communication such as books.

The confession, self-response, and response from an other or others constitutes a religious ritual that arises to meet the needs of at least some everyday environmentalists: the need to deal with the moral and existential questions that arise when one fails to embody one's values and live up to one's own moral standards (Fredericks 2014b). To return to Bell's language, they develop a "rite of affliction," here to resolve a problem the agent perceives to have brought upon themselves (Bell 2009b, 115). Here, the affliction is one that individuals perceive to have brought upon themselves. By writing their confessions, everyday environmentalists articulate their ideals; recognize how their actions differ from their ideals; and identify, reflect upon and, in this case, verbally act out how things ought to be according to their worldview with others in their community, a key component of J. Z. Smith's idea of ritual (Smith 1980, 123–35; 1987, 41, 109). Indeed, participants often articulate a vision of how human limits and possibility can coexist to address the existential issues raised by people who continually fail to live up to their own standards (Fredericks 2014b).

In these exchanges, participants fulfill many of the conditions necessary to respond to guilt and shame: they embody, reinforce, and normalize their values and visions of what it means to be human, forming a group or collective with shared concerns and actions. They build trust and credibility among participants such that advice and moral support can increase over time. Environmental confessional rituals create the participants' individual and collective identity and shape expectations about how environmental limitations are to be dealt with—that is, primarily through future confessions though also through changed actions.

These spontaneous confessional rituals, however, typically lack or fail to fully represent several conditions. Many environmental confessions focus primarily on the individual, leaving participants vulnerable to narcissism. Explicit ritualization to give more room for self-reflection and connection to others may stem this tendency. Acknowledging the collective nature of many environmental wrongs and our state of being would also enable more effective responses to guilt and shame. Then rituals would respond to the proper level of agency and recognize the relationality of the situation. Furthermore, addressing the collective as well as individual nature of environmental guilt and shame can enable participants to comprehensively deal with their environmental guilt and shame rather than merely assuaging guilt in the near future. This is not to say that short-term emotional needs are not important, but rather that fully addressing emotional and existential needs will require a more sustained effort. These confessional rituals are also limited insofar as the confessants and respondents are the only agents involved and both groups are generally comprised of people who have done wrong or are themselves considered bad. There is no contact with the wronged—to recognize them, to learn how to help them, or let alone to apologize or participate in restorative justice. Thus, the rituals are not as relational and thus not as effective as they need to be to deal with the many dimensions of guilt and shame.

Granted, some confessional processes, particularly those revealed in books and longer online discussions are more likely to pay explicit attention to these issues. This suggests that sustained attention to environmental guilt and shame may enable more robust rituals. One way to facilitate such depth is to deliberately shape confessional rituals, using experts from within the community of participants to ensure that the revised rituals resonate with the community. Experts can help participants name and recognize the relational aspects of guilt and shame: that the individuals, members, and collectives directly harm others by degrading their environment and continuing to do so even when they recognize what they are doing, causing psychological harm to their victims. Expert intervention should steer action not only to promote the most direct physically effective solutions, e.g. promoting alternative energy or minimizing flying or to those that provide quick emotional relief, of say fixating on a recycling strategy as *the* answer to environmental guilt, but also to encourage systems of apology and reconciliation, i.e. rituals between and among individuals and collectives that address psychological, relational, physical, and environmental harms. A hurdle to enhancing such efforts in existing online environmental confessional spaces is that confessions are made to the void of the internet in general or to a community largely comprised of other concerned environmentalists without leadership sufficiently reflective about the nuances of guilt and shame to recognize the interconnections of individual and collective action, danger of quick and easy assuaging of guilt, or the depth of damage shame can bring.

While these spontaneously developing rituals meet many of the conditions of responding to environmental guilt and shame, they leave room for improvement with respect to several conditions: identification with the other and relationality with respect to building relationships with those harmed. Without communication with those affected by climate change, culturally, economically, and physically appropriate responses may not be possible because confessors are likely to focus on their own emotions and criteria of how to respond to wrongdoing. To address all the dimensions of the guilt related to harming other people, agents must acknowledge the physical and economic elements of such harm as well as the relational, emotional, and cultural aspects. For instance, to address the harms of contributing to climate change by using coal fired power from coal mined using mountaintop removal techniques, agents should recognize multiple negative outcomes: the pain of no longer being able to fish safely in now polluted streams, relationships to humans and other biota that are degraded as landscapes change and traditional ways of life are disrupted, and the distrust of and anger toward perpetrators who ignore the devastating effects of their actions. Apologies are one way to partly address these harms, but they are not possible in current formulations of online confessions where the parties involved are typically not those harmed by anthropogenic environmental degradation (Fredericks 2019). Rather, improved relationships between perpetrator and victim would depend on the unlikely possibility that a victim would stumble across the confession and building a relationship with the confessor. Consequently, the confessional ritual will often benefit from targeting specific interlocutors, even if such targets are a subset of those harmed. This will require more intentionality, leadership, and community building than we have seen so far in the internet rituals. Intentional leadership, even by informal leaders, can identify a particularly relevant audience for the confession and helping foster relationships between confessants and victims. In this way, confessions could better facilitate positive relational, physical, psychological, social, and economic knowledge and outcomes that require participation from all affected.

One example of such targeted apologies occurred at the Dakota Access Pipeline (DAPL) prayer camp at the Standing Rock Sioux Reservation. Specifically, the December 5, 2016 ceremony of confession, apology, and request for forgiveness initiated by some veterans for collective harms to Native Americans by the U.S. government, particularly the military. While the veteran's actions were far from perfect and certainly do not deserve credit for the success of the movement, they illustrate many important features of ritual action stemming from collective guilt and shame and begin to address some conditions neglected by the online confessions. Thus, after describing the background to the apology ceremony, my analysis centers on the veterans developing and participating in this ritual.

In 2016, increased opposition to the DAPL became one of the most famous environmental movements in recent U.S. history: the NODAPL movement, particularly a prayer camp started by Native youth at the Standing Rock Sioux Reservation in North Dakota. Activism centered on the decision to route the pipeline just north and upstream of the Reservation and under the Missouri river, the most contentious aspect of the pipeline which runs over 1000 miles from North Dakota to Illinois (Thornbecke 2016). The prayer camp was a place for ceremonial activities of purity, cleansing, support for youth, and preservation of culture and land (Elbein 2017; Estes 2016). Organizers of this and other camps that sprung up were committed to peaceful resistance without alcohol, drugs, or weapons (Wehelie).

Many Native Americans, as well as nonnative environmental activists, and civil rights leaders argued that moving away from a previously considered route closer to Bismarck perpetuated long-standing injustice toward the Sioux (Elbein 2017; Thornbecke 2016). The risk of a contaminated water supply was one more problem, along with alcoholism, abuse, and lack of opportunities faced by Native communities due to settler colonialism (Elbein 2017).[4] Additionally, the pipeline siting process was expedited without full consultation from locals or attention to environmental and cultural impacts, a violation of treaty rights and the peoples' individual rights.[5]

Youth called attention to these issues among Indigenous Americans and the broader public (Elbein 2017). Eventually the movement of "water keepers" gained support from all of the Lakota bands, a historic partnership that had not happened "in modern memory" (Elbein 2017; Valen Levinson 2017). By May, 80,000 people had signed a petition to stop the pipeline (Revesz 2016). As attention to the issue grew throughout the summer and fall, more people joined the camps, swelling their numbers to several thousand (Thornbecke 2016). Members of other tribes in the United States, Indigenous people around the world, participants in the Black Lives Matter movement, representatives from many environmental organizations and religions, elected officials, veterans, and numerous celebrities joined the Lakota in person (García 2017; NODAPL; Taliman 2016a). Hundreds of thousands more wrote or called the Army Corps of Engineers, elected officials, and banks financing the pipeline (Hildreth). Marches and protests in over three hundred cities worldwide also aimed to stop the pipeline, reroute it, or at least convince the Army Corps of Engineers to conduct a full environmental and cultural analysis of the planned route (Democracy Now! 2016). Participants were concerned about Indigenous rights, risks to waterways for all sorts of creatures, and/or the fact that oil transported by the pipeline would exacerbate climate change.

While the vast majority of actions at the camp were nonviolent, as requested by the Standing Rock Sioux and in keeping with their vision of prayer or ceremonial camps (Clark and Wood 2016; Elbein 2017; Tribe 2017; Yates 2017), the water protectors faced threats and physical violence from DAPL security and law enforcement officials. They were bitten by guard dogs, teargassed, and sprayed

[4] For a discussion of the settler state as it relates to fossil fuel use and environmental degradation including climate change see (Whyte 2017).

[5] When the Army Corps has decision-making power over a pipeline it, as a part of the federal government, must also follow the National Historic Preservation Act which requires considering the effect of a new project on historical sites (Colwell 2016).

with water cannons in freezing weather (Hersher 2017; Sammon 2016; Thornbecke 2016). Beyond these direct bodily harms, "DAPL workers bulldozed irreplaceable sacred sites and burial grounds after court documents revealed their locations" (Taliman 2016a).

Reports of this treatment increased attention to and support for the Standing Rock. For instance, Wesley Clark Jr., a U.S. Army veteran, grew concerned when a Native friend of a friend called him and asked for assistance (Linehand 2016). After trying to help from afar with little result, Clark decided that he should lead a group of veterans to Standing Rock (van Gelder 2016). With Michael Wood Jr., a friend, veteran, and "advocate for national police reform," he reached out to veterans and raised money to offset their travel costs (Linehand 2016).[6]

As is the case for many supporters of the NODAPL movement, Clark expressed multiple motivations for his participation. Often, he called such service "his Christian duty" (Linehand 2016) or referenced general religious reasons for his service (Taliman 2016a). He also seemed to link the possibility of a spiritual call to serve to the healing he sought for his PTSD. Additionally, he aimed to protect peoples' Constitutional rights which he thought were violated by the pipeline construction and violence against water protectors (Taliman 2016a; van Gelder 2016). Many veterans who traveled to Standing Rock reported sharing at least some of Clark's motivations. Others sought a sense of purpose after leaving the military; wanted to continue living up to their pledge to protect the United States; or acted out of respect for Native Americans who frequently serve in the military and who have been wronged by the U.S. government in the past. Several veterans mentioned their concern for nature, particularly sparked by environmental degradation of places important to their childhood, as a catalyst for their participation (Basl 2017; Gabbard 2016; Linehand 2016; Taliman 2016a; 2016b; van Gelder 2016).

Legal and physical threats in the weeks before Clark and Wood's visit to Standing Rock also seem to have spurred the veterans' participation. On November 25, the Army Corps of Engineers wrote to the leaders of the Standing Rock Sioux indicating that as of December 5 it would "close federal property [...] including the [...] Oceti Sakowin camp." After this time, campers "would be considered trespassing and subject to prosecution" (Taliman 2016a). Despite the fact that the Army Corps softened its stance a bit by noting that "it would not force anyone to move" many in and outside the camps feared that the attempt to close the camps would result in direct violence and disrupted supply chains (Taliman 2016a).

The veterans led by Clark intended to act as physical and symbolic shields between the water protectors and law enforcement officials. Many thought that

[6] Both Native and non-Native veterans had been active supporters of the NODAPL movement long before Clark and Wood became involved (Schrader 2016; Standing Rock Veterans 2016; Taliman 2016a; Yates 2017), but I will refer to their group using the shorthand "the veterans."

officials would be less likely to use force against veterans in uniform (Gabbard 2016; Linehand 2016; Taliman 2016a; Tolan 2016). Clark and Wood's plan emphasized the peaceful intent of their mission (Clark and Wood 2016; Linehand 2016). Nonetheless, some participants at the camps worried that the veterans would incite rather than quell violence or that those with PTSD would not handle the sometimes tense situation well (Linehand 2016; Valen Levinson 2017; van Gelder 2016; Yates 2017).

The group led by Clark and Wood arrived on the weekend of December 3 and 4, 2016, growing to around 4000 people, double what they expected (Basl 2017; Grady 2016; Taliman 2016a). Surprisingly, on December 4, the Army Corps of Engineers announced it would not grant the easement for the DAPL to go under the Missouri River.[7] Celebrations quickly moved throughout the camp (Wong 2016). While the veterans' plans to act as human shields no longer made sense, other aspects of Clark and Wood's plan could still be pursued.

The predistributed timeline of events for December 5 included a "dawn welcoming ceremony" followed by "a cleansing ceremony to wash away our sins and steel us for the days ahead" (Clark and Wood 2016). The apology and forgiveness portion of the event was at least partly anticipated by Lakota and non-Native veterans alike. Some Lakota including Faith Spotted Eagle maintained that "the veterans think they're coming to protect us" "but they're not coming to protect us. They're coming to be healed" (Yates 2017), an observation reflecting the self-reported aims of multiple veterans who went to Standing Rock in part to be healed or to seek forgiveness (Basl 2017; Linehand 2016; Taliman 2016a).

While Clark had hoped for a military-style formation to address the Native Americans, a blizzard forced an indoor ceremony (Taliman 2016a; Tolan 2016). Drawing from written accounts (Amatulli 2016; Yates 2017) and video clips (Schrader 2016; wcmctv 2016), the ritual appears to have been directed by the Lakota (as followed the general leadership structure of the camp) with some forethought, though the events were not well practiced as indicated by the uncertain movements of some participants, especially veterans.

Near the beginning of the ceremony, at least one Native American Elder, Faith Spotted Eagle spoke, inviting participants to speak in "this safe space" noting that she "look[ed] forward to all of this making of relationship with you from these precious days that we have..." (Amatulli 2016; wcmctv 2016).[8] As she spoke, Clark raised his hand. She beckoned him to the front of a cluster of about a dozen

[7] When President Trump took office in January, 2017, he issued a memorandum to support the pipeline's construction (Trump 2017). The Army Corps stopped its environmental review, the pipeline was completed, and oil began to flow despite continued ceremonies and legal battles (Associated Press 2017; EarthJustice 2017). Legal battles continued during pipeline leaks and initiatives to increase the capacity of the pipeline (EarthJustice 2020; Nauman 2019).

[8] Youtube clips are not necessarily a full record of the events; a complete account of the ceremony has not yet been published, to my knowledge.

veterans, some in uniform. He was wearing the “blue jacket and hat of the 19th century 7th Calvary, evoking the 140-year-old memory of Gen. George Armstrong Custer” (Tolan 2016). After a word or two that is garbled in the videos, he clearly stated

> We came here to be the conscience of the nation and when that conscience we must first confess our sins to you because many of us, me particularly, are from the units that have hunted you over the many years. We came. We fought you. We took your land. We signed treaties that we broke. We stole minerals from your sacred hills. We blasted the faces of our presidents onto your sacred mountain. And we took still more land. And then we took your children. And then we tried to take your language. And we tried to eliminate YOUR language that God gave YOU and that the creator gave YOU. We didn’t respect you. We polluted your earth. We’ve hurt you in SO many ways. And we’ve come to say that we are sorry. (Amatulli 2016; wcmctv 2016)

At this point, Clark went down on one knee. He continued “We are at your service and we beg for your forgiveness.” Others veterans also kneeled. It seems that off camera, “Leonard Crow Dog, a Lakota elder and highly-regarded activist” motioned Clark forward. Clark scooted forward, still on one knee as he removed his hat. Crow Dog placed his outstretched hand on Clark’s lowered head (Amatulli 2016). After several seconds, he removed his hand. Clark backed up. They nodded to each other. Ululations could be heard from the crowd. Clark made a move as if to stand, but then went back down to one knee with a bowed head and stayed down for some time.

Ivan Looking Horse, a veteran, made some motions of direction and then began leading people in song (Amatulli 2016; wcmctv 2016). Crow Dog recalled the beginning of the world and said he would “accept the forgiveness [pause] world peace.” Next he discussed the longstanding and continued Lakota nation, their language, their service as caretakers, and how they treat the land differently than the general U.S. American population (Oceti Sakowin Camp 2016). He urged people (specifically the leaders of the United Nations and the Foreign Relations Committee) to recognize the law of the Lakota, to recognize his people as human beings, and to honor the Lakota nation. Referring to past treaties, he spoke of how Lakota have been treated, how they wish to be recognized, and how they intend to act in the future. Then Clark and Crow Dog hugged (Oceti Sakowin Oceti Sakowin Camp 2016). The ceremony continued with more speeches from Lakota and handshakes and hugs among non-Native veterans and Native Americans (Amatulli 2016).

Photographs and videos of the ceremony show that many participants were visibly moved by the ritual as faces crumpled, they fought to maintain their composure, wiped away their tears, engaged in long handshakes, or held each other in deep hugs. News interviews and first-hand accounts confirm the meaningful nature of the ceremony for people of many backgrounds (Abulhawa 2016; Oceti Sakowin Camp 2016; Taliman 2016b; Yates 2017).

Written sources and video evidence indicate about 500 people were present at the ceremony. The horrible weather, poor communication, or a lack of interest likely contributed to the fact that only a fraction of the veterans at Standing Rock participated in the ritual.[9] Yet, the ceremony itself and the widespread media attention to it indicate something about the importance of ritual for dealing with guilt and shame about environmental (and other) collective acts. As we have seen throughout this book, some people feel a need to apologize for what their collective has or continues to do. Clark's choice to wear the uniform of Custer's men at the Battle of the Little Big Horn indicates his identification with historic and ongoing collectives whose past actions constitute their continuing identity and shape the identity, responsibility, guilt, and shame of those who have become part of the collective recently (Tolan 2016). His word choice reinforces this identification. In his speech he only referred to himself as an individual once. The rest of the time he used a collective "we" referring to the United States, various branches of the military, or specific units (Amatulli 2016; wcmctv 2016). He did not blame the harms to the Lakota on bad leadership or leaders making role-based decisions, but rather considered the problems to arise and be sustained through a system of actions—military, government, societal—of which he and his companions were a part because of their military legacy and experience as U.S. Americans. The harms were recognized as long-term collective acts arising out of a collective ethos targeting a collective, its concepts, and way of life. Thus, the apologies have a collective nature even if they are not from the whole national collective or its official representatives.

Despite identifying with the historic and contemporary perpetrators of harm to the Lakota, Clark recognized that those assembled were a slightly different cohort, the self-proclaimed "conscience of the nation" (Amatulli 2016; wcmctv 2016). This smaller collective believed it understood what is right and when action to preserve justice is required. Clark wanted to shape the ideals and actions of the nation by serving as an inspirational, pivotal example. Before and at Standing Rock he claimed that the events there would be "most important event up to this time in human history" (Linehand 2016). He asserted: "I can promise you that in the next 12 months we are going to destroy the hydrocarbon industry.

[9] Some thought that an apology would blame people who are not to blame for the sins of their ancestors (Linehand 2016). Adam Linehan of *Task and Purpose*, a news site for and by veterans, objected to the ceremony on principle, because he claimed that the forgiveness ceremony was never mentioned in plan distributed to the veterans, and because it was a surprise for veterans at Standing Rock. The predistributed plan did not use the word "apology," so he may have thought it was a general cleansing ritual rather than specific to past harms perpetrated by the U.S. and the military in particular against the Lakota. Clark and Woods' well-documented poor communication at the camp likely contributed to this confusion.

I don't mean trying to damage anything—we are just going to tell people the truth about what it's doing to them" (Taliman 2016b). With these quotes, we see Clark's naïveté and overconfident presumptuousness. It is as if no one else had protested fossil fuels or sought to educate the public about their dangers. Yet, these bold statements also reveal his aim to spread his message of apology, support for the Lakota, and cessation of fossil fuel use. He wanted to call not just veterans but the whole nation to a new understanding of responsibility and honor.

Events leading up to the apology ritual and the event itself did begin to form a new collective with different ideals and practices as well as new relationships with the Lakota, as Spotted Eagle recognized (wcmctv 2016). Indeed, the apologies were only a subset of the veterans' intended actions (Gabbard 2016; Taliman 2016a; Yates 2017). Clark pledged to return to Standing Rock to help (Taliman 2016b), and described his desire to aid other people whose rights were being violated, especially with respect to access to safe water, particularly in Flint, MI (Linehand 2016; van Gelder 2016). The Veterans Stand organization, founded by Wood, and apparently an elaboration of the Veterans Stand with Standing Rock movement, aimed to "inspire, change, and empower communities through art, education, and service" (Veterans Stand 2017). They envisioned activities to "provide a platform for veterans to continue their service at home" and art events featuring artists from Standing Rock to "bring people together through empathy and shared experience."

Had these other activities come to pass, then Clark, Wood, and other veterans feeling shame could have given and received moral support, taken transformative action, avoided navel-gazing narcissism, and continued the process of becoming different agents. Yet, the potential of their initiatives was limited from the beginning. They did not sufficiently plan for the physical needs of veterans in their group or set up adequate means of communication. These logistical limitations, exacerbated by a blizzard, stranded about a thousand people at the Standing Rock Casino. Many veterans noted that they did not receive their promised reimbursements for travel. Veterans and some Native Americans blamed Clark and Wood for mismanaging financial and material resources at the camp (Abulhawa 2016; Linehand 2016; Tolan 2016; Unicorn Riot 2016). Clark, and to some degree his group at large, were also criticized for stealing the spotlight from the Standing Rock Sioux and the longer-term water protectors who had been at the camp for days, weeks, or months before their arrival (Schrader 2016; Standing Rock Veterans 2016; Taliman 2016b; Yates 2017). Eventually Clark was either run out of the veterans group or asked to step down (Linehand 2016). As far as I can tell, Veterans Stand with Standing Rock did not exist much, if at all, past December 2016. Its many dreams were not realized.

Given these multiple problems, this veteran apology at Standing Rock is far from an exemplary model of how to deal with collective guilt or shame. A one-time ceremony without a larger set of commitments and ongoing relationships to maintain transformations of self-conception and action will not address guilt or shame at anything but a cursory level. While individual participants may have changed their actions and identity to some degree, there is no evidence that the collective sustained the ongoing practices of acknowledgement, reconciliation, and behavioral change needed to address the depths of the wrongs they committed or the moral emotions they experience (Fredericks 2019, 145–56).

Yet this ritual, flawed as it is, can contribute to our study of ritual responses to environmental guilt and shame. It corroborates my claims that at least some people living in the United States feel the need for rituals to address both the wrongs giving rise to environmental guilt and shame and these emotions themselves. The veterans seemed to recognize the dignity of the Standing Rock Sioux as they responded to the Sioux's multiple requests for aid from Indigenous and non-Indigenous people.[10] Generally the aid seemed to be given and accepted on the terms of the Native Americans who were in charge of the ceremony camp.[11] For instance, the veterans changed their plans to avoid direct action on the weekend of December 3rd and followed the lead of Native elders during the ceremony. These modes of conduct indicate that many non-Natives were trying to recognize the personhood, authority, and values of the Lakota. Similarly, many Native Americans recognized the veterans' humanity, as demonstrated by the Natives' willingness to engage with them in general. A clear example was Spotted Eagle's compassionate recognition of the veterans' PTSD from the perspective of a community who also experienced it from "wars in this country" (Schrader 2016). Insofar as the apology ritual at the prayer camp recognized the dignity of others, particularly those harmed by environmental and social degradation, it complements (1) online confessional rituals which emphasize perpetrators and (2) ecological restoration rituals which focus on restoring the land and on the perpetrators of environmental degradation. Significantly, the confession and apology ritual was delivered to some of the individuals and collective harmed, and could have been a significant step in forging new relationships, modes of action, and identity.

Yet, to embody the conditions set out in Chapter 8, the confession and apology ritual and surrounding events requires extension in several ways. First, to be most effective at responding to the guilt and shame and the problems that sparked the guilt and shame, this ritual needed to catalyze ongoing action to address past problems and work to prevent them in the future. After all, the challenges of settler colonialism, environmental injustice, and climate change are ongoing. Responding to these cultural, political, and environmental issues as well as the

[10] Of course, not all Standing Rock Sioux desired such assistance or were even opposed to the pipeline; communities are not homogenous (Ravitz 2016).

[11] Certainly it would have been appropriate to come adequately prepared in order to avoid stressing the resources of the established camps.

shame that can accompany them will take more than a paragraph-long statement of apology during a weekend trip. It would take repeated action to change and maintain changes to identity, to build relationships, and to collaboratively determine how to respond to the many dimensions of such issues.

Second, the actions of this event need to be expanded beyond the particular communities involved, i.e. these veterans and the Standing Rock Sioux. The harms done by the U.S. government and its citizens are much more substantial than the harms addressed in the apology ritual and involve a much larger collective. While focusing on a subset of problems and people enables traction, even addressing the wrongs of the DAPL siting (to the degree possible given loss of sacred sites) would require much more significant intervention. Next steps could include supporting Native rights regarding a wide variety of issues and working to ensure environmental and procedural rights of all people. Elders call for this type of action in their speeches when they request respect and honor for the Lakota, political recognition of their nation, and particular modes of interaction with the United States government (Oceti Sakowin Camp 2016). While there were signals that such endeavors were dreamed about, follow-through is critical.

Third, regardless of the communities involved, engagement should be enriched. Sustained action should be accompanied by humility and willingness to listen to and partner with others. Participants will need deep, prolonged exchanges in order to build relationships, identify and understand problems, and devise and enact appropriate responses. Rituals of sharing one's stories, experiences, and emotions may also help participants get to know each other and to understand their various contributions to, experiences of, and emotions about environmental degradation. If interlocutors are spatially distant from each other, online communication can avoid some of the monetary and environmental expense of physical travel, but should be undertaken with care to ensure that relationships are built and actions and identity changed beyond the virtual space.

Fourth, engagement with others will likely need to be accompanied by significant, sustained self-reflection and transformation. Ritual activities can express and respond to one's emotional experience and transform one's sense of self, as Macy articulates. Reflection on new knowledge about others, oneself, and one's new relationships may transform one's understanding of one's own place in the world. This may well lead to guilt or shame. Formal processes of apology may be desired after such exchanges. Apologies and reconciliations could be facilitated in the ritual space and may be critical to the transformation of identity and actions among perpetrators of environmental damage. Such reconciliations can help transform self-judging shame or denial into a renewed sense of dignity and hope based in mutual trust and collaboration. While such apologies may overtly be directed at a particular group experiencing environmental shame, these apologies may also symbolize the need to reconcile with all humans or even the world (Fredericks 2019). Paying attention to the broader impacts of such partnerships and sharing their results with a wider audience will help move past a merely symbolic global interaction.

Alongside these more conversational and relational elements should be concrete actions to respond to the environmental and social degradation in question. With respect to the DAPL, initiatives could aim to not only shut down the pipeline but also directly reduce other fossil fuel use or develop new or enforce existing energy policy. Deliberate ritualization can help contextualize the abstract actions one takes—voting, writing a letter to an elected official, adjusting the thermostat—by clarifying their meaning and making the effects of such actions in complex abstract systems more visible.[12] For instance, one may be reminded through graphics on the thermostat or electric bill of some people or biota harmed by its use. People may say a prayer of remembrance when adjusting the thermostat or paying the electric bill. Through such ritualized actions, individual and community identity can be transformed into an identity of one who is not only responsible, guilty, and/or shameful but also one who takes action. Ritual can mark the movement from paralyzing shame to hopeful guilt, as ritualization is not limited to—and does not even always include—rational reflection on identity, but always makes and enacts identity.

Of course, even if all of these suggestions and more were fully followed, many—probably most—harms due to anthropogenic environmental degradation would not be erased. The guilty and shameful as well as those harmed will always need to deal with particular harms as well as the capacity to harm that seems part of being human. We need visions of identity which make sense of the horror people can perpetrate and articulate the promise of limiting and responding to such horror. Such visions should inform rituals to enact identity and bring about change. Concerted, intentional effort in these directions could go a long way toward recognizing and respecting the dignity of all and transforming the shameful into people who not only do bad things and recognize themselves as agents in a state of shame but also recognize that they are not fully consumed by such shame. They do not have to be paralyzed and/or doomed to continue "business as usual" environmental degradation.

The rituals at Standing Rock and of online confessions together suggest how environmental rituals of affliction can begin to foster the conditions necessary to respond to guilt and shame. While both of these rituals require elaboration on multiple levels, placing them side by side shows how rituals can operate at a collective level, attend to human–human and human–land harms, and move toward restorative justice. They can foster trust and credibility between parties, provide explicit and implicit moral support, develop and sustain values as well as visions of identity, enhance identification with the other, and enable critical

[12] People should not emphasize the ritual meaning of actions to the degree they ignore the most effective activities for combatting climate change.

self-examination and transformation. They can occur through and foster participatory processes. More theoretical and practical work is needed to identify best practices and to responsibly instantiate such rituals through the genuine participation of affected community members and ensure that such practices and rituals have a chance of being physically, emotionally, economically, socially, relationally, ethically, and legally effective. Doing so can help ensure that rituals resonate with the situation and tradition of participants, increasing the chance of the ritual being implemented, meaningful, and influential.

9.4 Intentional Ritualization

In the preceding analysis of rituals online and at Standing Rock as well as in my study of theories of environmental rituals of affliction, I identify ways in which these rituals or ritual theory fall short of meeting the conditions articulated in Chapter 8, and suggest ways to intentionally amend them in order to do so. Given the number of conditions required to holistically deal with the complexities of environmental guilt and shame, it is unlikely that spontaneous ritualization would fully bring about these conditions. Even if it could, a community should not count on it given the challenges posed by guilt and especially shame. Thus, to respond to and deal with environmental guilt and shame, people should intentionally develop and perform rituals, not just be satisfied with those that spontaneously emerge.

Such a project, however, faces several key challenges. Most obvious is the fact that there may not be the leadership necessary for intentional ritualization or the kind of community structures that would support a multifaceted ritual plan. This challenge is particularly acute in diffuse collectives. Thus, I support the intentional recognition and development of collectives, particularly to respond to the challenges of diffuse collectives to respond to complicated issues such as climate change. The work of Tracy Lynn Isaacs and Elizabeth Cripps, who advocate for the development of intentional collectives to deal with climate change, is instructive here, though my notion of collectives is more expansive than either of theirs and all of our theories require more practical development (Cripps 2013, 58–77; Isaacs 2011, 144–55).

Initiatives of intentional ritualization to meet the conditions are also challenging due to the many ways the conditions and rituals could interact. Namely, *one* ritual may not sufficiently meet all the conditions. Multiple rituals may be needed. Whether such events are considered separate rituals or subparts of a larger ritual is insignificant if the conditions are met in a community.

Another challenge for intentional ritualization is that modified or new rituals must resonate with their practitioners or else they will be meaningless at best. Participatory processes of ritual development can facilitate the ability of rituals to address the challenges faced by a community and avoid the ethical and practical challenges of the imposition of rituals. Striking a balance between spontaneous ritualization and critical reflection on and modification of rituals is difficult, as is the need to have broad participation in ritual creation while facilitating explicit reflection on the community's values and the conditions needed to respond to guilt and shame.

Indeed, the conditions themselves may be subject to modification or emendation if people with significantly different values follow these suggestions. I worked to make this set of conditions broad enough to be applicable to many situations/cultures, but recognize that they may not fit all. Yet, even as I recognize they are open to revision, I maintain that they are necessary, particularly if one values dignity and justice. If people do not attend to these conditions, they will be unlikely to fully deal with guilt or shame, likely exacerbating their negative outcomes (and the problematic behavior leading to them) over time.

Addressing all of these challenges of intentional ritualization is beyond the scope of this particular project, in part because it requires much more theoretical and practical study and in part because it will need to be specific to the particular cases at hand.

Pondering intentional ritualization to deal with guilt and shame may also turn our minds back to the question of whether guilt and shame should be intentionally induced, because the conditions and rituals are relevant not only to responding to pre-existing guilt and shame but also to any attempt to intentionally induce them. As discussed in previous chapters, I am wary of intentionally inducing these moral emotions due to the possibility that such strategies can neglect or erode human dignity and can exacerbate the very social and environmental conditions one hoped to curtail by inducing these moral emotions. Yet, I also leave the door to intentional inducement open a crack, for I think it may be done responsibly in theory if all the conditions are met—though the conditions are so difficult to meet simultaneously as to practically limit if not eliminate the possibility of shaming. To meet the conditions, if one is able to do so, one would need ritual, likely multiple rituals, in order to induce the guilt or shame, have a community capable of inducing such moral emotions, and a community able to hold the guilty or shameful person in the community and rehabilitate them. To do this one needs a close community of support, as can be fostered by rituals. Working out the details of such indicators would depend to a great extent on the particularities of the situation and would, like intentional ritualization, require intentional leadership as well as participatory processes in order for the process to be ethical. This too requires more research and specificity beyond this project.

Though much more research and practical experimentation is needed to develop rituals to deal with environmental guilt and shame and the problematic actions that catalyze these emotions, I have shown that ritual can theoretically meet the conditions of responding to guilt and shame, that other scholars have recognized the power of ritual to respond to environmental guilt and shame, and that emerging confessional and apology rituals attest to the power of such practices in the lives of everyday environmentalists. Rituals have many advantages as means of dealing with guilt and shame, and some are particularly relevant. They can involve bodily movement, emotion, reflective analysis, and identity- and relationship-building and maintenance, many importantaspects of being human. They enable the simultaneous and interacting development, maintenance, and sustaining of individual and collective identity and action. Insofar as many aspects of environmental guilt and shame have collective components, collective responses are needed. Rituals can enable them. Rituals also facilitate layers of symbolic and concrete action, as is necessary to attend both to the tangible effects of environmental guilt and shame and to those which are distant in space and time, largely hidden from everyday view, or which stem from the intertwining acts of individuals and collectives. For all these reasons, we should not be surprised that everyday environmentalists, whether in online communities or at Standing Rock, are developing rituals to respond to their guilt and shame. As we continue to grapple with anthropogenic environmental degradation, we would do well to remember the emotional effects of and contributions to such degradation, as well as the rituals that can aid individual and collective responses.

Epilogue
Looking Back, Looking Forward
Lessons from Studying Environmental Guilt and Shame

Over the course of the last nine chapters I have explored many questions arising from observing environmental guilt and shame among everyday environmentalists: What do guilt and shame mean? What kinds of agents can experience them? Ought agents feel guilt or shame? Should guilt or shame be induced? How do we respond to guilt or shame that already exists? Having articulated answers to these questions, I can now ask "What key research questions remain?" and deepen my answer to the question of Chapter 6, "What are the ramifications of studying environmental guilt and shame for environmental ethics, environmental action, and ethics in general?"

Paying attention to laypeople's ethics can yield insights that, when developed, overcome limitations of existing academic ethical analysis of climate change. While other scholars of religion, ethics, and the environment document the ethics of various groups, they rarely use these ideas to address philosophical or axiological problems, particularly beyond the typical environmental conversations. On the other hand, climate scholars who explicitly consider how climate change can influence axiological and philosophical arguments rarely engage directly with the ethics of everyday people. In this book, however, I take the writing of everyday environmentalists as a serious starting point for ethical reflection. Thus, I am able to identify and develop a response to possible impediments to environmental action (guilt and shame) that can account not only for individual and well-defined collectives but also for diffuse collectives. This multilayered theory of responsibility better describes the experience of climate change than models based on rational individual aggregation. It points us toward new ways of responding to climate change: collective rituals that attend to all elements of human experience. What other insights might we be able to develop with critical analyses of the ethics of other lay groups?

Recognizing the environmental guilt and shame of everyday environmentalists also provides indirect evidence for the spread of environmental values writ large. After all, if agents do not have values that they break or fail to achieve, they will not feel guilt or shame. Future social science research could engage directly with these agents to determine what values are most pressing and motivating. Such research could contribute to studies about how many people are "environmentalists" and whether religions are "greening." Attending to shame in such work

Environmental Guilt and Shame: Signals of Individual and Collective Responsibility and the Need for Ritual Responses.
Sarah E. Fredericks, Oxford University Press (2021). © Sarah E. Fredericks. DOI: 10.1093/oso/9780198842699.003.0010

would also help scholars recognize that deep commitment to environmental values may not lead to pro-environmental action. Changing ideals may not be enough to change action.

Acknowledging the collective perspective of many everyday environmentalists suggests that the metaphysics of interconnection may be more widespread and have greater implications than often thought. Environmental scholars often maintain that an environmental worldview is more holistic and interconnected than alternative modern Western worldviews but most analyses focuses on human-nature holism rather than the possibility of presuming human collectives. My research raises questions about the ways these lived worldviews do or do not relate to academic trends such as process thought and new materialism.

Environmental Guilt and Shame also contributes to environmental justice research in at least three ways and suggests new avenues for research related to each. First, I maintain that the association of individualism and environmental shame can put particular pressure on women and hinder environmental progress. More research is needed to uncover whether other demographic differences such as race and class affect the experience of environmental guilt and shame, understand whether such differences aid or hinder environmental action, and if so, ascertain how to combat such trends. Secondly, while my study examines environmental guilt and shame of people who are, globally speaking, environmental perpetrators, we should not presume that environmental guilt and shame are limited to such groups. Indeed, the small amount of research on this matter indicates that people in developing nations are more likely to report feeling environmental guilt and shame (National Geographic and GlobeScan 2012a, 77). Why? Do these feelings motivate or hinder environmental action? Third, as we consider the different experiences of environmental guilt and shame among various demographic groups, we should ask how we can build upon the suggestions of Chapter 9 to foster environmental and climate apology, restitution, adaptation, and mitigation. While some preliminary work has examined these issues, much more theoretical and practical work is needed here and on related concepts such as forgiveness (Almassi 2017a, 2017b; Fredericks 2019).

My study of guilt and shame also opens up possibilities in environmental ethics as it forces us to deal with human limits, complicated reactions to them, and the ramifications of such limits for our sense of self, emotions, self-judgment, relationships, actions, and our environmental conditions. All too often, environmental ethics, including (bizarrely) religious environmental ethics, has recognized that environmental problems have arisen because of limitations in our value systems, worldviews, policies, and habits, but then have posited revisions to values, worldviews, and policies as if such solutions will avoid fundamental human limits of knowledge, will, etc. Some authors, including Stephen Gardiner and Cynthia Moe-Lobeda, have begun to address these issues (Gardiner 2011;

Moe-Lobeda 2013, 2015), but much more attention to human capacities and limits is needed to develop ethics well suited not only to our ideals but also to the complications, tensions, and limitations of being human.

I hope that *Environmental Guilt and Shame* also indicates to my colleagues in environmental ethics, and any interested people in environmental movements or ethics writ large, that we do not need to fear negative emotions. Emphasizing positive emotions to avoid the stereotype of environmentalists as finger-wagging scolds, while well-intentioned, leaves some profound challenges unaddressed: the depths of emotions expressed by everyday environmentalists, the possibility that shame may exacerbate environmental degradation, and the disproportionate burden experienced by women while men and collectives are let off the hook. We can, as I hope to have demonstrated in this book, consider negative emotions including guilt and shame in order to determine when they are appropriate and how to deal with them without wallowing in them or inciting them at every turn. After all, people need ways to deal with negative emotions and their negative consequences. Since environmental guilt and shame will likely spread as environmental values spread and environmental degradation including climate change worsen, these needs will also increase. Rituals that ignore or only scratch the surface of guilt and shame can ultimately perpetuate shame cycles. Even if rituals cannot curb environmental degradation completely, they can set the stage for positive change if they break the shame spiral and promote a more mixed, layered, authentic view of agents.

Finally, given my attention to emotions and action through ritual, I push ethics toward engaging more comprehensively with what it means to be a human: not just our rationality but also our emotions, actions, emotions, relationships, embodiment, individual and collective identities. We see some moves in this direction in the turn to virtue ethics or ritual in existing literature, though it does not yet sufficiently attend to emotions. Ethics drawing on ritual activity integrates the development of ideals, the analysis of humans limits, and conceptual and practical means of responding to such failures. This turn to ritual also acknowledges the relational and often collective nature of our experience and identity as embodied beings who are affected by emotion, bodily movements, and relationships in general, and accepts that all of these dimensions are necessary when articulating and living by an ethic. A broader vision of ethics is needed to deal with all of the dimensions of environmental degradation—not just physical, economic, social, but also relational and emotional. In this way, the ethical approach I suggest by paying attention to emotions, including collective guilt and shame, and to rituals capable of responding to these complex emotions, challenges many prominent approaches to environmental ethics. Philosophical climate ethics, for instance, generally has focused on individual responsibility and linear causality such that it cannot account for collective responsibility for climate change, let alone the complexities of climate guilt and shame. Yet whenever a subset of our human experience is neglected in our ethics, such as emotions, fundamental finitude, or collective experience, our ethical ideals and practices will similarly be limited. We need ethical theories that are theoretically rigorous and implementable given the complexities and limitations of our world. My detailed study of environmental guilt and shame yields a multifaceted and integrated concept of ethics that can help us avoid the paralysis of environmental shame and leave us better prepared to respond to all sorts of ethical challenges, including and especially that of climate change.

References

Abulhawa, Susan. 2016. "Chronicle of Veterans at Standing Rock." veteranstoday.com. Accessed August 5, 2020. https://popularresistance.org/chronicle-of-veterans-at-standing-rock/.

Adam, Karla, and Faiz Siddiqui. 2015. "Climate Change Protests Take Place around the World on Eve of Summit." *The Washington Post*, November 29. Accessed October 6, 2019. https://www.washingtonpost.com/world/climate-change-protests-take-place-around-the-world-on-eve-of-summit/2015/11/29/9172b6ca-9530-11e5-befa-99cee-bcbb272_story.html.

Agence France-Presse. 2019. "Iceland Holds Funeral for First Glacier Lost to Climate Change." *The Guardian*, August 18.

Akehurst, Gary, Carolina Alfonso, and Helena Martins Goncalves. 2012. "Re-Examining Green Purchase Behaviour and the Green Consumer Profile: New Evidences." *Management Decision* 50(5): 972–88.

Albanese, Catherine L. 1990. *Nature Religion in America: From the Algonkian Indians to the New Age*. Chicago: University of Chicago Press.

Albanese, Catherine L. 2002. *Reconsidering Nature Religion*. Harrisburg, PA: Trinity Press International.

Almassi, Ben. 2017a. "Climate Change and the Need for Intergenerational Reparative Justice." *Journal of Agricultural & Environmental Ethics* 30(2): 199.

Almassi, Ben. 2017b. "Ecological Restorations as Practices of Moral Repair." *Ethics & the Environment* 22(1): 19–40.

Alpízar, Francisco, and Elisabeth Gsottbauer. 2015. "Reputation and Household Recycling Practices: Field Experiments in Costa Rica." *Ecological Economics* 120: 366–75.

Altintas, Evrim, and Oriel Sullivan. 2016. "Fifty Years of Change Updated: Cross-National Gender Convergence in Housework." *Demographic Research* 35: 455–70.

Amatulli, Jenna. 2016. "Forgiveness Ceremony Unites Veterans and Natives at Standing Rock Casino." *Huffington Post*. Accessed January 1, 2016. http://www.huffingtonpost.com/entry/forgiveness-ceremony-unites-veterans-and-natives-at-standing-rock-casino_us_5845cdbbe4b055b31398b199

Anderson, Christa M., Christopher B. Field, and Katharine J. Mach. 2017. "Forest Offsets Partner Climate-Change Mitigation with Conservation." *Frontiers in Ecology & the Environment* 15(7): 359–65.

Antonetti, Paolo, and Stan Maklan. 2014. "Feelings That Make a Difference: How Guilt and Pride Convince Consumers of the Effectiveness of Sustainable Consumption Choices." *Journal of Business Ethics* 124(1): 117–34.

Appiah, Anthony. 1991. "Racism and Moral Pollution." In *Collective Responsibility: Five Decades of Debate in Theoretical and Applied Ethics*, ed. Larry May and Stacey Hoffman, 219–38. Lanham, MD: Rowman & Littlefield.

Arendt, Hannah. 1991. "Organized Guilt and Universal Responsibility." In *Collective Responsibility: Five Decades of Debate in Theoretical and Applied Ethics*, ed. Larry May and Stacey Hoffman, 273–83. Lanham, MD: Rowman & Littlefield.

Arendt, Hannah. 1998. *The Human Condition*. Introduction by Margaret Canovan. Chicago: The University of Chicago Press.

Armisen, Fred, Carrie Brownstein, Karey Dornetto, and Jonathan Krisel. 2012. "Cool Wedding." *Portlandia* (Independent Film Channel).

Asma, Stephen T. 2010. "Green Guilt." [online] *The Chronicle of Higher Education*. Accessed February 25, 2020. http://chronicle.com/article/Green-Guilt/63447/.

Assmann, Aleida, and Anja Schwarz. 2013. "Memory, Migration and Guilt." *Crossings: Journal of Migration & Culture* 4(1): 51–65.

Associated Press. 2017. "A Timeline of the Dakota Access Pipeline." Accessed August 17. https://www.usnews.com/news/business/articles/2017-03-28/a-timeline-of-the-dakota-access-oil-pipeline.

Atkinson, Lucy. 2014. "Green Moms: The Social Construction of a Green Mothering Identity Via Environmental Advertising Appeals." *Consumption, Markets & Culture* 17(6): 553–72.

Bagdonas, James R. 2014. "Under Pressure." *Modern Family* (TV series).

Bagozzi, Richard P., Willem Verbeke, and Jacinto C. Gavino Jr. 2003. "Culture Moderates the Self-Regulation of Shame and Its Effects on Performance: The Case of Salespersons in the Netherlands and the Philippines." *Journal of Applied Psychology* 88(2): 219–33.

Bamberg, Sebastian, and Guido Möser. 2007. "Twenty Years after Hines, Hungerford, and Tomera: A New Meta-Analysis of Psycho-Social Determinants of Pro-Environmental Behaviour." *Journal of Environmental Psychology* 27(1): 14–25.

Banerjee, Neela. 2014. "Thousands Pack New York's Streets to March against Climate Change." *The Los Angeles Times*, September 21. Accessed October 4, 2019. https://www.latimes.com/nation/la-na-nn-climate-change-march-new-york-20140921-story.html.

Barbarossa, Camilla, and Patrick Pelsmacker. 2016. "Positive and Negative Antecedents of Purchasing Eco-Friendly Products: A Comparison between Green and Non-Green Consumers." *Journal of Business Ethics* 134(2): 229–47.

Barkan, Elazar. 2000. *The Guilt of Nations: Restitution and Negotiating Historical Injustices*. New York: W. W. Norton.

Barrett, Karen Caplovitz. 1995. "A Functionalist Approach to Shame and Guilt." In *Self-Conscious Emotions: The Psychology of Shame, Guilt, Embarrassment, and Pride*, ed. June Price Tangney and Kurt W. Fischer, 25–63. New York: The Guilford Press.

Barsade, Sigal G., and Andrew P. Knight. 2015. "Group Affect." *Annual Review of Organizational Psychology and Organizational Behavior* 2(1): 21–46.

Bartky, Sandra Lee. 1990. "Shame and Gender." In *Femininity and Domination: Studies in the Phenomenology of Oppression*, 83–98. New York: Routledge.

Basl, Kevin. 2017. "Kevin Basl: Why I, a Veteran, Joined the Fight at Standing Rock." *Winona Daily News*.

Bastiaansen, J. A. C. J., M. Thioux, and C. Keysers. 2009. "Evidence for Mirror Systems in Emotions." *Philosophical Transactions of the Royal Society B: Biological Sciences* 364(1528): 2391–2404.

Baugh, Amanda. 2016. *God and the Green Divide: Religious Environmentalism in Black and White*. Oakland: University of California Press.

Baumeister, Roy F., Arlene M. Stillwell, and Todd F. Heatherton. 1995. "Interpersonal Aspects of Guilt: Evidence from Narrative Studies." In *Self-Conscious Emotions: The Psychology of Shame, Guilt, Embarrassment, and Pride*, ed. June Price Tangney and Kurt W. Fischer, 255–73. New York: The Guilford Press.

BBC News. 2019. "What Is Flygskam? Greta Speaks up About 'Flight-Shaming'." BBC, July 19. https://www.bbc.co.uk/newsround/49032117.

Beavan, Colin. 2009. *No Impact Man: The Adventures of a Guilty Liberal Who Attempts to Save the Planet and the Discoveries He Makes About Himself and Our Way of Life in the Process*. New York: Farar, Straus and Giroux.

Bell, Catherine. 1997. *Ritual: Perspectives and Dimensions*. New York: Oxford University Press.

Bell, Catherine. 2009a. *Ritual Theory, Ritual Practice*. Oxford: Oxford University Press.

Bell, Catherine. 2009b. *Ritual: Perspectives and Dimensions*. New York: Oxford University Press.

Bellinger, David C., and Andrew M. Bellinger. 2006. "Childhood Lead Poisoning: The Torturous Path from Science to Policy." *Journal of Clinical Investigation* 116(4): 853–7.

Benedict, Ruth. 1967. *The Chrysanthemum and the Sword: Patterns of Japanese Culture*. Cleveland, OH: World Publishing Company.

Benford, Gregory. 1995. "Humanity as Cancer." *The Magazine of Fantasy and Science Fiction*, 95–102.

Bennett, Graceann, and Freya Williams. 2011. "Mainstream Green: Moving Sustainability from Niche to Normal." In *The Red Papers*. Chicago: Ogilvy & Mather.

Bennett, John W., and Michio Nagai. 1953. "Echoes: Reactions to American Anthropology." *American Anthropologist* 55(3): 404–11.

Berners-Lee, Mike. 2011. *How Bad Are Bananas? The Carbon Footprint of Everything*. Vancouver: Greystone Books.

Bingham, Sally. 2007. "Climate Change: A Moral Issue." In *Creating a Climate for Change: Communicating Climate Change and Facilitating Social Change*, ed. Susanne C. Moser and Lisa Dilling, 153–66. Cambridge: Cambridge University Press.

Bissing-Olson, Megan J., Kelly S. Fielding, and Aarti Iyer. 2016. "Experiences of Pride, Not Guilt, Predict Pro-Environmental Behavior When Pro-Environmental Descriptive Norms Are More Positive." *Journal of Environmental Psychology* 45: 145–53.

Blair, James, and Katherine Fowler. 2008. "Moral Emotions and Moral Reasoning from the Perspective of Affective Cognitive Neuroscience: A Selective Review." *European Journal of Developmental Science* 2(3): 303–23.

Bloomfield, Michael John. 2014. "Shame Campaigns and Environmental Justice: Corporate Shaming as Activist Strategy." *Environmental Politics* 23(2): 263–81.

Boden, Tom, and Bob Andres. 2014. *Ranking of the World's Countries by 2014 Total CO2 Emissions from Fossil-Fuel Burning, Cement Production, and Gas Flaring Carbon Dioxide*. Information Analysis Center, Oak Ridge National Laboratory; Research Institute for Environment, Energy and Economics, Appalachian State University.

Bogel-Burroughs, Nicholas, and Coral Davenport. 2019. "Climate Protesters Snarl Traffic but Washington Still Goes to Work." *The New York Times*, September 23. Accessed October 4, 2019. https://www.nytimes.com/2019/09/23/climate/dc-climate-protest.html.

Boles, Elson. 2006. "Ruth Benedict's Japan: The Benedictions of Imperialism." *Dialectical Anthropology* 30(1): 27–70.

Brandalism. 2015. Brandalism at Cop21//Paris//2015. Brandalism.org. Accessed March 23, 2021.

Branscombe, Nyla R., and Bertjan Doosje, eds. 2004a. *Collective Guilt: International Perspectives*. New York: Cambridge University Press.

Branscombe, Nyla R., and Bertjan Doosje. 2004b. "International Perspectives on the Experience of Collective Guilt." In *Collective Guilt: International Perspectives*, ed. Nyla R. Branscombe and Bertjan Doosje, 3–15. New York: Cambridge University Press.

Branscombe, Nyla R., Ben Slugoski, and Diane M. Kappen. 2004. "The Measurement of Collective Guilt: What It Is and What It Is Not." In *Collective Guilt: International Perspectives*, ed. Nyla R. Branscombe and Bertjan Doosje, 16–34. New York: Cambridge University Press.

Bray, Jeffery, Nick Johns, and David Kilburn. 2010. "An Exploratory Study into the Factors Impeding Ethical Consumption." *Journal of Business Ethics* 98(4): 597–608.

Brown, Rupert, Roberto González, Hanna Zagefka, Jorge Manzi, and Sabina Čehajić. 2008. "Nuestra Culpa: Collective Guilt and Shame as Predictors of Reparation for Historical Wrongdoing." *Journal of Personality and Social Psychology* 94(1): 75–90.

Buck, Dina. 2013. "When a Bookaholic Also Loves Forests: Trying to Reconcile the Environmental Footprint of Books." *Pachamama*, April 30. http://www.pachamama.org/blog/when-a-bookaholic-also-loves-forests-trying-to-reconcile-the-environmental-footprint-of-books.

Burnett, Melissa S., and Dale A. Lunsford. 1994. "Conceptualizing Guilt in the Consumer Decision-Making Process." *Journal of Consumer Marketing* 11(3): 33–43.

Byrka, Katarzyna, Terry Hartig, Florian G. Kaiser. 2010. "Environmental Attitude as a Mediator of the Relationship between in Nature and Self-Reported Ecological Behavior." *Psychological Reports* 107(3): 847–59.

Cafaro, Philip. 2015. "Three Ways to Think About the Sixth Mass Extinction." *Biological Conservation* 192: 387–93.

Callicott, J. Baird. 2014. *Thinking Like a Planet*. New York: Oxford University Press.

Callicott, J. Baird. 2017a. "Thinking Climate Change Like a Planet." In *Teaching Climate Change in the Humanities*, ed. Stephen Siperstein, Shane Hall and Stephanie LeMenager, 79–85. Abingdon: Routledge.

Callicott, J. Baird. 2017b. "Worldview Remediation in the First Century of the New Millennium." In *Environmental Humanities: Voices from the Anthropocene*, ed. Serpil Oppermann and Serenella Iovino, 133–53. Lanham, MD: Rowman & Littlefield.

Campbell, Polly.2010. "Going Green Without Guilt." *Sustainable Self*. https://selfsustain.com/blog/going-green-without-the-guilt-how-to-conquer-eco-guilt-while-reaching-your-green-potential/.

Campaign Against Climate Change. 2010. National Climate March 2010.

Chan, Ricky Y. K., and Lorett B. Y. Lau. 2002. "Explaining Green Purchasing Behavior." *Journal of International Consumer Marketing* 14(2–3): 9–40.

Chase, Elaine, Kyomuhendo, and Grace Bantebya, eds. 2015. *Poverty and Shame: Global Experiences*. Oxford: Oxford University Press.

Chea, Terence. 2006. "Part of the Solution or a License to Pollute?" *Chicago Sun-Times*, December 11.

Christov-Moore, Leonardo, and Marco Iacoboni. 2015. "Emotions in Interaction: Toward a Supraindividual Study of Empathy." In *Emotion in Group Decision and Negotiation*, ed. Bilyana Martinovsky, 1–32. Dordrecht: Springer.

Cisneros, Elías, Sophie Lian Zhou, and Jan Börner. 2015. "Naming and Shaming for Conservation: Evidence from the Brazilian Amazon." *PLoS ONE* 10(9).

Clark, Wesley Jr., and Michael Wood Jr. 2016. "Veterans Stand for Standing Rock Operations Order."

Clowney, David. 2013. "Biophilia as an Environmental Virtue." *Journal of Agricultural & Environmental Ethics* 26(5): 999–1014.

Cole, David. 2008. "Creative Sentencing: Using the Sentencing Provisions of the South Australian Environment Protection Act to Greater Community Benefit." *Environmental and Planning Law Journal* 25(1): 13–21.

Colwell, Chip. 2016. "How the Archaeological Review Behind the Dakota Access Pipeline Went Wrong." Accessed August 12, 2020. https://theconversation.com/how-the-archaeological-review-behind-the-dakota-access-pipeline-went-wrong-67815.

Cone Communications. 2007. "All That Glitters Isn't Gold: This Year, It's Green." Press release. Accessed January 20. https://www.csrwire.com/press_releases/19356-all-that-glitters-isn-t-gold-this-year-it-s-green

Confucius. 1992. *The Analects*. Translated by D. C. Lau. Hong Kong: Chinese University Press.
Conrad, Sarah M. 2015. "A Restorative Environmental Justice for the Prison Industrial Complex: A Transformative Feminist Theory of Justice." PhD, Philosophy and Religion Studies, University of North Texas.
Conradie, Ernst. 2008. *The Church and Climate Change*. Pietermaritzburg: Cluster Publications.
Cooper, David E. 1991. "Collective Responsibility (a Defense)." In *Collective Responsibility: Five Decades of Debate in Theoretical and Applied Ethics*, ed. Larry May and Stacey Hoffman, 35–46. Lanham, MD: Rowman & Littlefield.
Coulter, Robin Higie, and Mary Beth Pinto. 1995. "Guilt Appeals in Advertising: What Are Their Effects?" *Journal of Applied Psychology* 80(6): 697–705.
Creighton, Millie R. 1990. "Revisiting Shame and Guilt Cultures: A Forty-Year Pilgrimage." *Ethos* 18(3): 279–307.
Cripps, Elizabeth. 2013. *Climate Change and the Moral Agent: Individual Duties in an Interdependent World*. Oxford: Oxford University Press.
Crislip, Kathleen. 2012. "Carbon Offests on Planes and Green Travel Guilt: Or How I'm Glad I Learned to Love the Bus." Accessed February 25, 2020. http://studenttravel.about.com/od/greentravel/qt/Green-Travel-Guilt.htm.
Crutzen, Paul, and Will Steffen. 2003. "How Long Have We Been in the Anthropocene Era?" *Climate Change* 61: 251–7.
Crutzen, Paul J., and Eugune F. Stoermer. 2000. "The 'Anthropocene'." *IGBP Newsletter* 41.
Cunsolo, Ashlee, and Karen Landman, eds. 2017. *Mourning Nature : Hope at the Heart of Ecological Loss and Grief*. Montreal: McGill-Queen's University Press.
Dahl, Darren W., Heather Honea, and Rajesh V. Manchanda. 2003. "The Nature of Self-Reported Guilt in Consumption Contexts." *Marketing Letters* 14(3): 159–71.
Darkow, John. 2016. "Untitled." *The Funny Times*, April 9.
Davidson, Debra J., and William R. Freudenburg. 1996. "Gender and Environmental Risk Concerns: A Review and Analysis of Available Research." *Environment and Behavior* 28(3): 302–39.
de Klerk, Barend J. 2014. "Enhancing Ecological Consciousness through Liturgical Acts of Doxology and Lament." *Verbum et Ecclesia* 35(2): 1–8.
de Vries, David. 2008. *Life after People*. History Channel.
de Vries, David. 2009. *Life after People: The Complete Season One*. History Channel.
de Vries, David. 2010. *Life after People: The Complete Season Two*. History Channel.
Democracy Now! 2009. "Bolivian President Evo Morales: 'Shameful' for West to Spend Trillions on War and Only $10 Billion for Climate Change." *Democracy Now!* Accessed October 8, 2016. http://www.democracynow.org/2009/12/16/bolivian_president_evo_morales_shameful_for
Democracy Now! 2016. "Dakota Access Pipeline Protests Spread to 300 Cities as Pipeline Owner Sues to Continue Construction."
Dicum, Gregory. 2007. "Pleasure without Guilt: Green Hotels with Comfort." *New York Times* 157(54172): F1–4.
Dodd, Donald. 2012. "Green Guilt: How Bad Do You Feel—or Not—About Cleaning Your Plate?" *The Salem News Online*. Accessed January 20, 2014. http://www.thesalemnewsonline.com/opinion/publishers_column/donald_dodd/article_d88c6f88-01a9-11e2-9223-001a4bcf6878.html
Doherty, Thomas Joseph. 2015. "Sustainable Self." Sustainable Self, LLC. Accessed August 25, 2020. http://selfsustain.com/.
Dolezal, Luna. 2015. *The Body and Shame: Phenomenology, Feminism, and the Socially Shaped Body*. Lanham, MD: Lexington Books.

Doosje, Bertjan, Nyla R. Branscombe, Russell Spears, and Antony S. R. Manstead. 2004. "Consequences of National Ingroup Identification for Responses to Immoral Historical Events." In *Collective Guilt: International Perspectives*, ed. Nyla R. Branscombe and Bertjan Doosje, 95–111. New York: Cambridge University Press.

Dube, Rebecca. 2012. "Survey: Parents Feel 'Green Guilt' over Plastic Bags, Diapers." *TODAY Moms*. Accessed January 20, 2014. http://www.today.com/moms/survey-parents-feel-green-guilt-over-plastic-bags-diapers-718631

Dunbar-Ortiz, Roxanne. 2014. *An Indigenous People's History of the United States*. Boston, MA: Beacon Press.

EarthJustice. 2017. "The Standing Rock Sioux Tribe's Litigation on the Dakota Access Pipeline." EarthJustice. Accessed August 17, 2020. http://earthjustice.org/features/faq-standing-rock-litigation.

EarthJustice. 2020. "The Standing Rock Sioux Tribe's Litigation on the Dakota Access Pipeline." Accessed February 25, 2020. https://earthjustice.org/features/faq-standing-rock-litigation.

Edelstein, Robin S., and Philip R. Shaver. 2007. "A Cross-Cultural Examination of Lexical Studies of Self-Conscious Emotions." In *The Self-Conscious Emotions: Theory and Research*, ed. Jessica L. Tracy, Richard W. Robins, and June Price Tangney, 194–208. New York: The Guilford Press.

Ehrenreich, Barbara. 2001. *Nickled and Dimed: On (Not) Getting By in America*. New York: Henry Holt.

Eisner, Andrew. 2013. "Earth Day 2013 Special Gadgetology Report: Who Is Recycling Electronics and Buying Green Gadgets?" *Retrevo Blog*. http://www.retrevo.com/content/blog/2013/04/earth-day-2013-special-gadgetology-report.

Elbein, Saul. 2017. "The Seventh Generation." *The New York Times*, February 5, 24. https://www.nytimes.com/2017/01/31/magazine/the-youth-group-that-launched-a-movement-at-standing-rock.html.

Elgaaied, Leila. 2012. "Exploring the Role of Anticipated on Guilt on Pro-Environmental Behavior: A Suggested Typology of Residents in France Based on Their Recycling Patterns." *Journal of Consumer Marketing* 29(5): 369–77.

Epstein, Charlotte, and Kate Barclay. 2013. "Shaming to 'Green': Australia–Japan Relations and Whales and Tuna Compared." *International Relations of the Asia-Pacific* 13(1): 95–123.

Estes, Nick. 2016. "Fighting for Our Lives: #NODAPL in Historical Context." *therednation.org*, September 18. https://therednation.org/fighting-for-our-lives-nodapl-in-context/.

Eurich-Rascoe, Barbara L. 1997. *Femininity and Shame: Women, Men, and Giving Voice to the Feminine*. Lanham, MD: University Press of America.

Everett, Laura. 2017. *Holy Spokes: The Search for Urban Spirituality on Two Wheels*. Grand Rapids, MI: William B. Eerdmans.

Fahrenthold, David A. 2008. "There's a Gold Mine in Environmental Guilt." *The Washington Post*, October 6. Accessed April 16, 2012. http://www.washingtonpost.com/wp-dyn/content/article/2008/10/05/AR2008100502518_2.html.

Feinberg, Joel. 1991. "Collective Responsibility (a Reply to Cooper)." In *Collective Responsibility: Five Decades of Debate in Theoretical and Applied Ethics*, ed. Larry May and Stacey Hoffman, 53–76. Lanham, MD: Rowman & Littlefield.

Ferguson, Mark A., and Nyla R. Branscombe. 2010. "Collective Guilt Mediates the Effect of Beliefs About Global Warming on Willingness to Engage in Mitigation Behavior." *Journal of Environmental Psychology* 30(2): 135–42.

Ferguson, Mark A., Nyla R. Branscombe, and Katherine J. Reynolds. 2011. "The Effect of Intergroup Comparison on Willingness to Perform Sustainable Behavior." *Journal of Environmental Psychology* 31(4): 275–81.

Ferguson, Tamara J., Daniel Brugman, Jennifer White, and Heidi L. Eyre. 2007. "Shame and Guilt as Morally Warranted Experiences." In *The Self-Conscious Emotions: Theory and Research*, ed. Jessica L. Tracy, Richard W. Robins, and June Price Tangney, 330–48. New York: The Guilford Press.

Ferguson, Tamara J., and Hedy Stegge. 1995. "Emotional States and Traits in Children: The Case of Guilt and Shame." In *Self-Conscious Emotions: The Psychology of Shame, Guilt, Embarrassment, and Pride*, ed. June Price Tangney and Kurt W. Fischer, 174–97. New York: The Guilford Press.

Fessler, Daniel M. T. 2007. "From Appeasement to Conformity: Evolutionary and Cultural Perspectives on Shame, Competition, and Cooperation." In *The Self-Conscious Emotions: Theory and Research*, ed. Jessica L. Tracy, Richard W. Robins, and June Price Tangney, 174–93. New York: The Guilford Press.

Figueroa, Robert Melchior. 2006. "Evaluating Environmental Justice Claims." In *Forging Environmentalism: Justice, Livelihood, and Contested Environments*, ed. Joanne Bauer, 360–76. New York: M. E. Sharpe.

Figueroa, Robert Melchior, and Gordon Waitt. 2010. "Climb: Restorative Justice, Environmental Heritage, and the Moral Terrains of Uluru-Kata Tjuta National Park." *Environmental Philosophy* 7(2): 135–63.

Finnegan, Eleanor. 2011. "What Traditions Are Represented in Religion and Ecology? A Perspective from an American Scholar of Islam." In *Inherited Land: The Changing Grounds of Religion and Ecology*, ed. Whitney A. Bauman, Richard R. Bohannon II, and Kevin J. O'Brien, 64–79. Eugene, OR: Pickwick.

Fischer, Kurt W., and June Price Tangney. 1995. "Self-Conscious Emotions and the Affect Revolution: Framework and Overview." In *Self-Conscious Emotions: The Psychology of Shame, Guilt, Embarrassment, and Pride*, ed. June Price Tangney and Kurt W. Fischer, 3–22. New York: The Guilford Press.

Fisher, Berenice. 1984. "Guilt and Shame in the Women's Movement: The Radical Ideal of Action and Its Meaning for Feminist Intellectuals." *Feminist Studies* 10(2): 185–212.

Flesher, Jared. 2009. "A Decline in 'Green Guilt.'" *Green: A Blog About Energy and the Environment*. http://green.blogs.nytimes.com/2009/04/21/a-decline-in-green-guilt/.

Forencich, Frank. 1992. *Homo Carcinomicus: A Look at Planetary Oncology*. Edmonton: Athabasca University Press.

Foster, John. 2015. *After Sustainability: Denial, Hope, Retrieval*. Abingdon: Routledge.

Fourie, Melike M., Kevin G. F. Thomas, David M. Amodio, Christopher M. R. Warton, and Ernesta M. Meintjes. 2014. "Neural Correlates of Experienced Moral Emotion: An FMRI Investigation of Emotion in Response to Prejudice Feedback." *Social Neuroscience* 9(2): 203.

Fraj, Elena, and Eva Martinez. 2007. "Ecological Consumer Behaviour: An Empirical Analysis." *International Journal of Consumer Studies* 31(1): 26–33.

Fredericks, Sarah E. 2014a. *Measuring and Evaluating Sustainability: Ethics in Sustainability Indexes*. New York: Routledge.

Fredericks, Sarah E. 2014b. "Online Confessions of Eco-Guilt." *Journal for the Study of Religion, Nature and Culture* 8(1): 64–84.

Fredericks, Sarah E. 2016. "Ritual Responses to Climate Engineering." In *Theological and Ethical Reflections on Climate Engineering: Calming the Storm*, ed. Forrest Clingerman and Kevin J. O'Brien. Lanham, MD: Lexington Books.

Fredericks, Sarah E. 2019. "Climate Apology and Forgiveness." *Journal of the Society of Christian Ethics* 39(1): 143–59.

Freeman, Kyle. 2008. "Survey: Does Eco Guilt Stop You from Shopping at Ikea?" Apartment Therapy. Accessed January 20, 2020. http://www.apartmenttherapy.com/survey-does-eco-64103.

French, Peter A. 1979. "The Corporation as a Moral Person." *American Philosophical Quarterly* 16(3): 207–15.

Gabbard, Tulsi. 2016. "Thousands of Veterans Travel to Standing Rock to Support Activists". In *All Things Considered*, ed. Michel Martin. npr.org.

Gallese, Vittorio. 2001. "The 'Shared Manifold' Hypothesis: From Mirror Neurons to Empathy." *Journal of Consciousness Studies* 8(5–7): 33–50.

Gallese, Vittorio, and Alvin Goldman. 1998. "Mirror Neurons and the Simulation Theory of Mind-Reading." *Trends in Cognitive Sciences* 2(12): 493–501.

García, Michelle. 2017. "Standing Rock's Men at War." *High Country News*, 26–7.

Gardiner, Stephen M. 2011. *A Perfect Moral Storm: The Ethical Tragedy of Climate Change*. New York: Oxford University Press.

Gardiner, Stephen M. 2012. "Are We the Scum of the Earth? Climate Change, Geoengineering, and Humanity's Challenge." In *Ethical Adaptation to Climate Change: Human Virtues of the Future*, ed. Allen Thompson and Jeremy Bendik-Keymer, 182–96. Cambridge, MA: MIT Press.

Gardiner, Stephen M. 2017. "Accepting Collective Responsibility for the Future." *Journal of Practical Ethics* 5(1): 22–52.

Glassberg, Brooke Kosofsky. 2008. "Getting over Eco-Guilt". *O, The Oprah Magazine* (May).

Geertz, Clifford. 1973. *The Interpretation of Cultures*. New York: Basic Books.

Gerber, Lisa. 2002. "What Is So Bad About Misanthropy?" *Environmental Ethics* 24(1): 41–55.

Gibson, David. 2009. "Awash in Green: A Critical Perspective on Environmental Advertising." *Tulane Environmental Law Journal* 22(2): 423–40.

Gilbert, Daniel. 2006. "We Are the Villain." *Los Angeles Times*, July 2. https://search-proquest-com.proxy.uchicago.edu/docview/422079494?accountid=14657.

Gilbert, Paul. 2007. "The Evolution of Shame as a Marker for Relationship Security: A Biopsychosocial Approach." In *The Self-Conscious Emotions: Theory and Research*, ed. Jessica L. Tracy, Richard W. Robins, and June Price Tangney, 283–309. New York: The Guilford Press.

Gilead, Michael, Maayan Katzir, Tal Eyal, and Nira Liberman. 2016. "Neural Correlates of Processing 'Self-Conscious' vs. 'Basic' Emotions." *Neuropsychologia* 81: 207–18.

Gneezy, Uri, and Aldo Rustichini. 2000. "A Fine Is a Price." *Journal of Legal Studies* 29(1): 1–17.

Goetz, Jennifer L., and Dacher Keltner. 2007. "Shifting Meanings of Self-Conscious Emotions across Cultures: A Social-Functional Approach." In *The Self-Conscious Emotions: Theory and Research*, ed. Jessica L. Tracy, Richard W. Robins, and June Price Tangney, 153–73. New York: The Guilford Press.

Gore, Albert. 2006. *An Inconvenient Truth: The Planetary Emergency of Global Warming and What We Can Do About It*. Emmaus, PA: Rodale.

Gottlieb, Robert. 1993. *Forcing the Spring: The Transformation of the American Environmental Movement*. Washington, DC: Island Press.

Grady, J. 2016. *12/6/16 Commander Update Wes Clark Jr*. YouTube.

Graton, Aurélien, François Ric, and Emilie Gonzalez. 2016. "Reparation or Reactance? The Influence of Guilt on Reaction to Persuasive Communication." *Journal of Experimental Social Psychology* 62: 40–9.

Gray, John. 2002. *Straw Dogs: Thoughts on Humans and Other Animals*. Cambridge: Granta.

Gray, Louise. 2013. "David Attenborough: Humans Are Plague on Earth." *Telegraph*. Accessed September 10. http://www.telegraph.co.uk/news/earth/earthnews/9815862/Humans-are-plague-on-Earth-Attenborough.html

Greene, Danielle, Parisa Tehranifar, Diana P. DeMartini, Andrew Faciano, and Deborah Nagin. 2015. "Peeling Lead Paint Turns into Poisonous Dust. Guess Where It Ends Up? A Media Campaign to Prevent Childhood Lead Poisoning in New York City." *Health Education & Behavior* 42(3): 409.

Gruenwald, Tara L., Sally S. Dickerson, and Margaret E. Kemeny. 2007. "A Social Function for Self-Conscious Emotions: The Social Self Preservation Theory." In *The Self-Conscious Emotions: Theory and Research*, ed. Jessica L. Tracy, Richard W. Robins, and June Price Tangney, 68–87. New York: The Guilford Press.

Gwen. 2009a. "Clean Slate." *Green with Envy, Riddled with Guilt*. https://greenguilt.wordpress.com/2009/01/05/clean-slate/.

Gwen. 2009b. "*My Agenda*." *Green with Envy, Riddled with Guilt*. https://greenguilt.wordpress.com/my-agenda/.

Haber, Ben. 2008. "*Is Green Making You Blue?*" *Race Talk Blog*. http://racetalkblog.com/2008/05/08/is-green-making-you-blue/.

Hamilton, William. 2008. "I Hated How They Acted So Much 'Greener Than Thou.'" *The New Yorker*, February 11 and 18, 125.

Han, Heesup. 2014. "The Norm Activation Model and Theory-Broadening: Individuals' Decision-Making on Environmentally-Responsible Convention Attendance." *Journal of Environmental Psychology* 40: 462–71.

Hansen Lagattuta, Kristin, and Ross A. Thompson. 2007. "The Development of Self-Conscious Emotions: Cognitive Processes and Social Influences." In *The Self-Conscious Emotions: Theory and Research*, ed. Jessica L. Tracy, Richard W. Robins, and June Price Tangney, 91–113. New York: The Guilford Press.

Harder, David W. 1995. "Shame and Guilt Assessment, and Relationships of Shame- and Guilt-Proneness to Psychpathology." In *Self-Conscious Emotions: The Psychology of Shame, Guilt, Embarrassment, and Pride*, ed. June Price Tangney and Kurt W. Fischer, 368–92. New York: The Guilford Press.

Harding, Matthew, and David Rapson. 2019. "Does Absolution Promote Sin? A Conservationist's Dilemma." *Environmental and Resource Economics* 73(3): 923–55.

Harding, Robyn. 2009. *Mom, Will This Chicken Give Me Man Boobs?: My Confused, Guilt-Ridden and Stressful Struggle to Raise a Green Family*. Vancouver: Greystone Books.

Hart, Daniel, and M. Kyle Matsuba. 2007. "The Development of Pride and Moral Life." In *The Self-Conscious Emotions: Theory and Research*, ed. Jessica L. Tracy, Richard W. Robins, and June Price Tangney, 114–33. New York: The Guilford Press.

Hart, Joseph. 2008. "Guilty Liberals: Do Your Best, But Don't Burn Out on Shame." Utine Reader. Accessed January 20, 2020. http://www.utne.com/politics/guilty-liberals.aspx.

Harth, Nicole Syringa, Colin Wayne Leach, and Thomas Kessler. 2013. "Guilt, Anger, and Pride about In-Group Environmental Behaviour: Different Emotions Predict Distinct Intentions." *Journal of Environmental Psychology* 34: 18–26.

Hartmann, Patrick, and Vanessa Apaolaza-Ibanez. 2012. "Consumer Attitude and Purchase Intention toward Green Energy Brands: The Roles of Psychological Benefits and Environmental Concern." *Journal of Business Research* 65(9): 1254–63.

Hatfield, Elaine, John T. Cacioppo, and Richard L. Rapson. 1994. *Emotional Contagion*. Cambridge: Cambridge University Press.

Hatfield, Elaine, Megan Carpenter, and Richard L. Rapson. 2014. "Emotional Contagion as a Precursor to Collective Emotions." In *Collective Emotions: Perspectives from Psychology,*

Philosophy, and Sociology, ed. Christian von Scheve and Mikko Salmela, 108–22. Oxford: Oxford University Press.

Head, Lesley. 2016. *Hope and Grief in the Anthropocene: Re-Conceptualising Human-Nature Relations, Routledge Research in the Anthropocene*. Abingdon: Routledge.

Held, Virginia. 1991. "Can a Random Collection of Individuals Be Morally Responsible?" In *Collective Responsibility: Five Decades of Debate in Theoretical and Applied Ethics*, ed. Larry May and Stacey Hoffman, 89–100. Lanham, MD: Rowman & Littlefield.

Helm, Bennett W. 2014. "Emotional Communities of Respect." In *Collective Emotions: Perspectives from Psychology, Philosophy, and Sociology*, ed. Christian von Scheve and Mikko Salmela, 47–60. Oxford: Oxford University Press.

Helmer, Jodi. 2005. "Feel Guilty When You're Not Green?" Home-Dzine. Accessed February 25, 2020. http://www.home-dzine.co.za/Lifestyle/life-guilty.htm.

Hequembourg, Amy L., and Ronda L. Dearing. 2013. "Exploring Shame, Guilt, and Risky Substance Use among Sexual Minority Men and Women." *Journal of Homosexuality* 60(4): 615.

Hern, Warren M. 1993a. "Has the Human Species Become a Cancer on the Planet? A Theoretical View of Population Growth as a Sign of Pathology." *Current World Leaders* 36(6): 1.

Hern, Warren M. 1993b. "Is Human Culture Carcinogenic for Uncontrolled Population Growth and Ecological Destruction?" *BioScience* 43(11): 768–73.

Hersher, Rebecca. 2017. "Key Moments in the Dakota Pipeline Fight." *NPR*, February 22. http://www.npr.org/sections/thetwo-way/2017/02/22/514988040/key-moments-in-the-dakota-access-pipeline-fight.

HeyAllie. 2011. "Make Me Actually Use My Reusable Bags!" [online discussion board] Accessed February 25, 2020. http://ask.metafilter.com/187308/Make-me-actually-use-my-reusable-bags.

Hickman, Leo. 2005. *A Life Stripped Bare: My Year Trying to Live Ethically*. London: Transworld.

Hickok, Gregory. 2014. *The Myth of Mirror Neurons: The Real Neuroscience of Communication and Cognition*. New York: W. W. Norton.

Hildreth, Matthew A. "Stop the Violence against the Standing Rock Sioux Tribe." Accessed February 25, 2020. https://sign.moveon.org/petitions/tell-the-state-of-north.

Hines, Jody M., Harold R. Hungerford, and Audrey N. Tomera. 1986/87. "Analysis and Synthesis of Research on Responsible Environmental Behavior: A Meta-Analysis." *Journal of Environmental Education* 18(2): 1–8.

Hochschild, Arlie Russell. 1990. *The Second Shift*. New York: Avon Books.

Hopkins, Michael S., and Catherine Roche. 2009. "What the 'Green' Consumer Wants." *MIT Sloan Management Review* 50(4): 87.

Howe, Jeff. 2005. "I'm Feeling Guilty—and I'm Not Catholic." [online] Dovetail Partners. Accessed February 25, 2020. http://www.dovetailinc.org/content/im-feeling-guilty-and-im-not-catholic.

Isaacs, Tracy Lynn. 2011. *Moral Responsibility in Collective Contexts*. Oxford: Oxford University Press.

Isaacs, Tracy. 2013. "Response to Critics." *Dialogue* 53(01): 43–55.

Isaacs, Tracy. 2014. "Collective Responsibility and Collective Obligation." *Midwest Studies in Philosophy* 38: 40–57.

Iwasawa, Tomoko. 2012. "Philosophical Faith as the Will to Communicate: Two Case Studies in Intercultural Understanding." In *Philosophical Faith and the Future of Humanity*, 347–63. New York: Springer.

Jacquet, Jennifer. 2015a. *Is Shame Necessary?: New Uses for an Old Tool*. New York: Pantheon Books.
Jacquet, Jennifer. 2015b. "Why Guilt and Ethical Shopping Aren't Enough—We Need to Start Shaming." *The Guardian*, February 6. http://www.theguardian.com/books/2015/feb/06/point-of-view-why-shame-make-world-better-place.
Jacquet, Jennifer, and George Dyson. 2011. "Is Shame Necessary?" *Edge*. Accessed June 15, 2016. https://www.edge.org/conversation/jennifer_jacquet-is-shame-necessary
Jaggar, Alison M. 1989. "Love and Knowledge: Emotion in Feminist Epistemology." *Inquiry: An Interdisciplinary Journal of Philosophy* 32(2): 151–76.
James, George Alfred. 2013. *Ecology Is Permanent Economy: The Activism and Environmental Philosophy of Sunderlal Bahuguna*. Albany, NY: SUNY.
James, Simon P. 2014. "'Nothing Truly Wild Is Unclean': Muir, Misanthropy, and the Aesthetics of Dirt." *Environmental Ethics* 36(3): 357–63.
Jamieson, Dale. 1996. "Ethics and Intentional Climate Change." *Climatic Change* 33: 323–36.
Jamieson, Dale. 2014. *Reason in a Dark Time: Why the Struggle against Climate Change Failed—and What It Means for Our Future*. New York: Oxford University Press.
Jamieson, Dale. 2015. "Responsibility and Climate Change." *Global Justice: Theory Practice Rhetoric* 8(2): 23–42.
Jaspers, Karl. 1969. *Tragedy Is Not Enough*. Translated by Harald A. T. Reiche, Harry T. Moore, and Karl W. Deutsch. Hamden, CT: Archon Books.
Jaspers, Karl. 1970. *Philosophy*. Translated by E. B. Ashton, vol. 2. Chicago: The University of Chicago Press.
Jaspers, Karl. 2001. *The Question of German Guilt*. Translated by Joseph W. Koterski SJ. New York: Fordham University Press.
Jaspers, Karl. 2002. "Reading Kierkegaard 2. Guilt: The Fundamental Limit Situation." In *Karl Jaspers on Philosophy of History and History of Philosophy*, ed. Joseph W. Koterski and Raymond J. Langley, 252–7. Amherst, NY: Humanity Books.
Jenkins, Willis. 2013. *The Future of Ethics: Sustainability, Social Justice, and Religious Creativity*. Washington, DC: Georgetown University Press.
Jenkins, Willis. 2014. "Atmospheric Powers, Global Injustice, and Moral Incompetence: Challenges to Doing Social Ethics from Below." *Journal of the Society of Christian Ethics* 34(1): 65–82.
Jimenez, Marissa, and Kenneth C. C. Yang. 2008. "How Guilt Level Affects Green Advertising Effectiveness." *Journal of Creative Communications* 3: 231–54.
Johnson, Thomas. 2011. "Environmental Information Disclosure in China: Policy Developments and NGO Responses." *Policy and Politics* 39(3): 399–416.
Johnston, Lucas F. 2013. *Religion and Sustainability: Social Movements and the Politics of the Environment*. Bristol, CT: Equinox.
Johnstone, Gerry, and Daniel W. Van Ness. 2007. "Introduction." In *The Handbook of Restorative Justice*, ed. Gerry Johnstone and Daniel W. Van Ness, 1–4. Portland, OR: Willan.
Jones, Sarah. 2013a. "Green Guilt." *Green at Heart*. http://greenatheart.org/green-guilt/.
Jones, Sarah. 2013b. "Why Should We Live a Green Lifestyle?" *A Simple Green Life*, October 4. http://simplegreenlifestyle.com/green-lifestyle/.
Jones, Warren H., Karen Kugler, and Patricia Adams. 1995. "You Always Hurt the One You Love: Guilt and Transgressions against Relationship Partners." In *Self-Conscious Emotions: The Psychology of Shame, Guilt, Embarrassment, and Pride*, ed. June Price Tangney and Kurt W. Fischer, 301–21. New York: The Guilford Press.

Jordan, William R. III. 2003. *The Sunflower Forest: Ecological Restoration and the New Communion with Nature*. Berkeley, CA: University of California Press.

Jordan, William R. III. 1993. "Rituals of Restoration." *The Humanist* 53(6): 23–6.

Jordan, William R. III, Nathaniel F. Barrett, and Kip Curtis. 2012. "Foundations of Conduct: A Theory of Values and Its Implications for Environmentalism." *Environmental Ethics* 34(3): 291–312.

Joshi, Yatish, and Zillur Rahman. 2015. "Factors Affecting Green Purchase Behaviour and Future Research Directions." *International Strategic Management Review* 3(1–2): 128–43.

Joy. 2012. "To Market, to Market." *Frock Files*. http://frockfiles.com/2012/03/20/to-market-to-market/.

Kaiser, Florian G., P. Wesley Schultz, Jaime Berenguer, Victor Corral-Verdugo, and Geetika Tankha. 2008. "Extending Planned Environmentalism: Anticipated Guilt and Embarrassment across Cultures." *European Psychologist* 13(4): 288–97.

Kaiser, Florian G., and Todd A. Shimoda. 1999. "Responsibility as a Predictor of Ecological Behaviour." *Journal of Environmental Psychology* 19(3): 243–53.

Kamau, Caroline, Roger Giner-Sorolla, and Sven Zebel. 2013. "Reconciliation Responses, Blame, and Expressions of Guilt or Shame." *Journal of Applied Social Psychology* 43: E287–92.

Kaplan, Sarah, and Brady Dennis. 2019. "'We Will Make Them Hear Us': Millions of Youths around the World Strike for Action." *The Washington Post*, September 20.

Kaplan, Temma. 2002. "Acts of Testimony: Reversing the Shame and Gendering the Memory." *Signs: Journal of Women in Culture and Society* 28(1): 179–99.

Katchadourian, Herant. 2010. *Guilt: The Bite of Conscience*. Stanford, CA: Stanford University Press.

Katzenstein, Suzanne. 2013. "Reverse-Rhetorical Entrapment: Naming and Shaming as a Two-Way Street." *Vanderbilt Journal of Transnational Law* 46: 1079.

Kellert, Stephen R., and Edward O. Wilson. 1993. *The Biophilia Hypothesis*. Washington, DC: Island Press.

Kent, Pauline. 1999. "Japanese Perceptions of the Chrysanthemum and the Sword." *Dialectical Anthropology* 24(2): 181–92.

Kitayama, Shinobu, Hazel Rose Markus, and Hisaya Matsumoto. 1995. "Culture, Self, and Emotion: A Cultural Perspective on 'Self-Conscious' Emotions." In *Self-Conscious Emotions: The Psychology of Shame, Guilt, Embarrassment, and Pride*, ed. June Price Tangney and Kurt W. Fischer, 439–65. New York: The Guilford Press.

Klein, Margaret. 2015. "The Climate Psychologist." Accessed September 8, 2020. http://theclimatepsychologist.com/about-the-author/.

Klein, Margaret. 2020. "The Climate Mobilization." Accessed March 7, 2020. https://www.theclimatemobilization.org/overview-and-team.

Knight, Les. "Voluntary Human Extinction Movement." Accessed September 3, 2020. http://www.vhemt.org/index.htm.

Koch, Wendy. 2009. "Going Green, Women Show More Eco-Guilt in New Year's Resolutions." *USA Today*, December 30. http://content.usatoday.com/communities/greenhouse/post/2009/12/going-green-women-show-more-eco-guilt-in-new-years-resolutions/1#.Ut1ifRDnaM-.

Koenig-Lewis, Nicole, Adrian Palmer, Janine Dermody, and Andreas Urbye. 2014. "Consumers' Evaluations of Ecological Packaging: Rational and Emotional Approaches." *Journal of Environmental Psychology* 37: 94–105.

Kornblum, Janet. 2008. "Survey: More Americans Feeling 'Green Guilt.'" *USA Today*, May 7. http://www.usatoday.com/news/nation/environment/2008-05-07-green-guilt_N.htm?loc=interstitialskip.

La Rovere, Luca. 2008. "The 'Examination of Conscience' of the Nation: The Lost Debate About the 'Collective Guilt' in Italy, 1943–5." *Totalitarian Movements & Political Religions* 9(2/3): 187–202.
Lack, Evonne. 2013. "Top 7 Mommy Guilt Trips—and How to Handle Them." Accessed January 20, 2020. http://www.babycenter.com/0_top-7-mommy-guilt-trips-8211-and-how-to-handle-them_3654967.bc?page=2#articlesection3.
Lasas, Ainius. 2008. "Guilt, Sympathy, and Cooperation: EU–Baltic Relations in the Early 1990s." *East European Politics & Societies* 22(2): 347–72.
Laseu, Dana-Nicoleta. 1991. "Consumer Guilt: Examing the Potential of a New Marketing Construct." *Advances in Consumer Research* 18: 290–5.
Lehtinen, Ullaliina. 1998. "How Does One Know What Shame Is? Epistemology, Emotions, and Forms of Life in Juxtaposition." *Hypatia* 13(1): 56–77.
Lertzman, Renee. 2015. *Environmental Melancholia: Psychoanalytic Dimensions of Engagement*. Abingdon: Routledge.
LeVasseur, Todd. 2011. "Shame, Ritual and Beauty: Technologies of Encountering the Other—Past, Present and Future." In *Placing Nature on the Borders of Religion, Philosophy, and Ethics*, ed. Mark H. Dixon and Forrest Clingerman, 25–44. Abingdon: Routledge.
Levinson, Adam Valen. 2017. "DAPL: The Push and Pull of Standing Rock." Al Jazeera (Qatar). https://www.aljazeera.com/features/2017/1/25/dapl-the-push-and-pull-of-standing-rock.
Lewis & Clark Public Relations Department. 2010. "Ecopsychology Earns National Media Attention." Lewis & Clark College. Accessed August 25, 2015. https://www.lclark.edu/live/news/4260-ecopsychology-earns-national-media-attention.
Lexmark International, Inc. 2009. "Consumers Concerned, Yet Clueless, on 'Green' Technology." Lexmark International. Accessed January 18, 2020. http://newsroom.lexmark.com/index.php?s=13630&item=23708.
Li, Jin, Lianqin Wang, and Kurt W. Fischer. 2004. "The Organisation of Chinese Shame Concepts." *Cognition & Emotion* 18(6): 767–97.
Lickel, Brian, Toni Schmader, and Marchelle Barquissau. 2004. "The Evocation of Moral Emotions in Intergroup Contexts: The Distinction between Collective Guilt and Collective Shame." In *Collective Guilt: International Perspectives*, ed. Nyla R. Branscombe and Bertjan Doosje, 35–55. New York: Cambridge University Press.
Lickel, Brian, Toni Schmader, and Marija Spanovic. 2007. "Group-Conscious Emotions: The Implications of Others' Wrongdoings for Identity and Relationships." In *The Self-Conscious Emotions: Theory and Research*, ed. Jessica L. Tracy, Richard W. Robins, and June Price Tangney, 351–70. New York: The Guilford Press.
Lindsay-Hartz, Janice, Joseph deRivera, and Michael F. Mascolo. 1995. "Differentiating Guilt and Shame and Their Effects on Motivation." In *Self-Conscious Emotions: The Psychology of Shame, Guilt, Embarrassment, and Pride*, ed. June Price Tangney and Kurt W. Fischer, 274–300. New York: The Guilford Press.
Lindsey, Anne-Marie. 2012. "[Pre]Maternal Guilt, Part Two: Green Guilt." *Do Not Faint*. http://www.psychologytoday.com/blog/do-not-faint/201202/pre-maternal-guilt-part-two-green-guilt.
Linehand, Adam. 2016. "Why They Went: The Inside Story of the Standing Rock Veterans." *Task & Purpose*. Accessed August 5, 2017. http://taskandpurpose.com/went-inside-story-standing-rock-veterans/
Liu, Jingru, Xin Sun, Bin Lu, Yunkun Zhang, and Rui Sun. 2016. "The Life Cycle Rebound Effect of Air-Conditioner Consumption in China." *Applied Energy* 184: 1026–32.
Locke, Jill. 2007. "Shame and the Future of Feminism." *Hypatia* 22(4): 146–62.

Macdonald, Doune, Rebecca Abbott, and David Jenkins. 2012. "Physical Activity of Remote Indigenous Australian Women: A Postcolonial Analysis of Lifestyle." *Leisure Sciences* 34(1): 39–54.
MacDougall, Kent. 1996. "Humans as Cancer." *Wild Earth* 6(3): 81–8.
MacKendrick, Norah A. 2010. "Media Framing of Body Burdens: Precautionary Consumption and the Individualization of Risk." *Sociological Inquiry* 80(1): 126–49.
Mackie, Diane M., and Eliot R. Smith. 2015. "Intergroup Emotions." In *Apa Handbook of Personality and Social Psychology*, vol. 2: *Group Processes*, ed. M. Mikulincer, P. R. Shaver, J. F. Dovidio and J. A. Simpson, 263–93. Washington, DC: American Psychological Association.
Macy, Joanna, and Molly Brown. 2014. *Coming Back to Life: The Updated Guide to the Work That Reconnects*. Gabriola Island: New Society.
Macy, Joanna, and Chris Johnstone. 2012. *Active Hope: How to Face the Mess We're In Without Going Crazy*. Novato, CA: New World Library.
Maddison, Sarah. 2012. "Postcolonial Guilt and National Identity: Historical Injustice and the Australian Settler State." *Social Identities* 18(6): 695–709.
Mallett, Robyn K., and Janet K. Swim. 2004. "Collective Guilt in the United States: Predicting Support for Social Policies That Alleviate Social Injustice." In *Collective Guilt: International Perspectives*, ed. Nyla R. Branscombe and Bertjan Doosje, 56–74. New York: Cambridge University Press.
Manion, Jennifer C. 2002. "The Moral Relevance of Shame." *American Philosophical Quarterly* 39(1): 73–90.
Manion, Jennifer C. 2003. "Girls Blush, Sometimes: Gender, Moral Agency, and the Problem of Shame." *Hypatia* 18(3): 21–41.
Markowitz, Gerald, and David Rosner. 2000. "'Cater to the Children': The Role of the Lead Industry in a Public Health Tragedy, 1900–1955." *American Journal of Public Health* 90(1): 36–46.
Mascolo, Michael F., and Kurt W. Fischer. 1995. "Developmental Transformations in Appraisals for Pride, Shame, and Guilt." In *Self-Conscious Emotions: The Psychology of Shame, Guilt, Embarrassment, and Pride*, ed. June Price Tangney and Kurt W. Fischer, 64–113. New York: The Guilford Press.
Mason, Michelle. 2010. "On Shamelessness." *Philosophical Papers* 39(3): 401–25.
Matthews, H. Damon. 2016. "Quantifying Historical Carbon and Climate Debts among Nations." *Nature Climate Change* 6(1): 60.
May, Larry. 1987. *The Morality of Groups: Collective Responsibility, Group-Based Harm, and Corporate Rights*. Notre Dame, IN: University of Notre Dame Press.
May, Larry. 1991. "Metaphysical Guilt and Moral Taint." In *Collective Responsibility: Five Decades of Debate in Theoretical and Applied Ethics*, ed. Larry May and Stacey Hoffman, 239–54. Lanham, MD: Rowman & Littlefield.
May, Larry. 1992. *Sharing Responsibility*. Chicago: The University of Chicago Press.
May, Larry. 1996. *The Socially Responsible Self: Social Theory and Professional Ethics*. Chicago: The University of Chicago Press.
May, Larry, and Stacey Hoffman, eds. 1991. *Collective Responsibility: Five Decades of Debate in Theoretical and Applied Ethics*. Lanham, MD: Rowman & Littlefield.
Mazar, Nina, and Chen-Bo Zhong. 2010. "Do Green Products Make Us Better People?" *Psychological Science* 21(4): 494–8.
McCann, Daryl. 2011. "The Personal Costs of Spurning Green Misanthropy." *Quadrant Magazine* 55(5): 34–7.

McCright, Aaron M., and Chenyang Xiao. 2014. "Gender and Environmental Concern: Insights from Recent Work and for Future Research." *Society & Natural Resources* 27(10): 1109–13.

McFarland Taylor, Sarah. 2009. *Green Sisters: A Spriritual Ecology*. Cambridge, MA: Harvard University Press.

McGarty, Craig, and Ana-Marie Bliuc. 2004. "Refining the Meaning of the 'Collective' in Collective Guilt: Harm, Guilt, and Apology in Australia." In *Collective Guilt: International Perspectives*, ed. Nyla R. Branscombe and Bertjan Doosje, 112–29. New York: Cambridge University Press.

McGary, Howard. 1991. "Morality and Collective Liability." In *Collective Responsibility: Five Decades of Debate in Theoretical and Applied Ethics*, ed. Larry May and Stacey Hoffman, 77–87. Lanham, MD: Rowman & Littlefield.

McKibben, Bill. 2016. "My Right-Wing Stalkers." *New York Times*, August 7, 4–5. http://proxy.uchicago.edu/login?url=http://search.ebscohost.com/login.aspx?direct=true&db=nfh&AN=117285649&site=eds-live&scope=site.

Mellers, B. A., and A. P. McGraw. 2001. "Anticipated Emotions as Guides to Choice." *Current Directions in Psychological Science* 10(6): 210–14.

Mencius. 2009. *Mencius*. Translated by Irene Bloom. New York: Columbia University Press.

Merchant, Carolyn. 1980. *The Death of Nature: Women, Ecology, and the Scientific Revolution*. New York: Harper & Row.

Miller, Rowland S. 1995. "Embarrassment and Social Behavior." In *Self-Conscious Emotions: The Psychology of Shame, Guilt, Embarrassment, and Pride*, ed. June Price Tangney and Kurt W. Fischer, 322–39. New York: The Guilford Press.

Moe-Lobeda, Cynthia D. 2013. *Resisting Structural Evil: Love as Ecological-Economic Vocation*. Minneapolis, MN: Fortress Press.

Moe-Lobeda, Cynthia. 2015. "Climate Debt, White Privilege and Christian Ethics as Political Theology." In *Common Goods: Economy, Ecology, and Political Theology*, ed. Catherine Kelleret al., 286–304. New York: Fordham University Press.

Moe-Lobeda, Cynthia. 2016a. "A Haunting Contradiction, Hope and Moral-Spiritual Power." In *Eco-Reformation: Grace and Hope for a Planet in Peril*, ed. Jim B. Martin-Schramm, Bill McKibben, and Lisa E. Dahill, 32–40. Eugene, OR: Wipf & Stock.

Moe-Lobeda, Cynthia D. 2016b. "Climate Change as Climate Debt: Forging a Just Future." *Journal of the Society of Christian Ethics* 36(1): 27–49.

Moe-Lobeda, Cynthia. 2017a. "Climate Injustice, Truth-Telling, and Hope." *Anglican Theological Review* 99(3): 531–40.

Moe-Lobeda, Cynthia. 2017b. "From Climate Debt to Climate Justice: God's Love Embodied in Garden Earth." In *The Wiley Blackwell Companion to Religion and Ecology*, ed. John Hart, 203–19. Hoboken, NJ: John Wiley.

Moon, Sangkil, Paul K. Bergey, Liliana L. Bove, and Stefanie Robinson. 2016. "Message Framing and Individual Traits in Adopting Innovative, Sustainable Products (Isps): Evidence from Biofuel Adoption." *Journal of Business Research* 69(9): 3553–60.

Mooney, Maryellen, and Gina Stikes. 2012. "Ivillage and Today.Com 'Eco-Confessions' Survey, Most Parents Admit They Could Be Greener." Accessed January 20, 2020. http://www.nbcnews.com/id/47064299/ns/about-press_releases/#.Ut2KwxDnaM9.

Morrow, James K. 1995. *Towing Jehovah*. New York: Harcourt.

Mueller, P. S. 2018. Shame Bins Cartoon. Cleveland Heights, OH: *Funny Times*.

Murdie, Amanda, and Johannes Urpelainen. 2015. "Why Pick on Us? Environmental INGOs and State Shaming as a Strategic Substitute." *Political Studies* 63(2): 353–72.

Nakagawa, Seishu, et al. 2015. "Comprehensive Neural Networks for Guilty Feelings in Young Adults." *NeuroImage* 105: 248–56.

Nash, James A. 1991. *Loving Nature: Ecological Integrity and Christian Responsibility*. Nashville, TN: Abingdon Press.

National Geographic. 2008. *Greendex 20098: Consumer Choice and the Environment: A Worldwide Tracking Survey*. Washington, DC: *National Geographic*.

National Geographic. 2009. *Greendex 2009: Consumer Choice and the Environment: A Worldwide Tracking Survey: Highlights Report*. Washington, DC: *National Geographic*.

National Geographic. 2010. *Greendex 2010: Consumer Choice and the Environment: A Worldwide Tracking Survey*. Washington, DC: *National Geographic*.

National Geographic and GlobeScan. 2012a. *Greendex 2012: Consumer Choice and the Environment: A Worldwide Tracking Survey*. Washington, DC: *National Geographic* and GlobeScan.

National Geographic and GlobeScan. 2012b. *Greendex 2012: Consumer Choice and the Environment: A Worldwide Tracking Survey: Highlights Report*. Washington, D.C.: *National Geographic*.

Nauman, Talli. 2019. "Standing Rock Seeks Hearing to Deny DAPL Bid to Double Oil Flow." indianz.com.

Nelson, Robert H. 2011. "*Living Through the End of Nature: The Future of American Environmentalism* by Paul Wapner." *Perspectives on Politics* 9(1): 150–2.

Nemeth, Darlyne G., Robert B. Hamilton, and Judy Kuriansky, eds. 2012. *Living in an Environmentally Traumatized World: Healing Ourselves and Our Planet*. Santa Barbara, CA: Prager.

Niedenthal, Paula M., and Markus Brauer. 2012. "Social Functionality of Human Emotion." *Annual Review of Psychology* 63(1): 259–85.

NODAPL. 2017. "Standing Rock Water Protectors." Accessed August 12, 2020. http://www.nodaplarchive.com/groups-and-celebrities.html.

Nordhaus, Ted, and Michael Shellenberger. 2007. *Break Through: From the Death of Environmentalism to the Politics of Possibility*. Boston, MA: Houghton Mifflin.

Norgaard, Kari Marie. 2011. *Living in Denial: Climate Change, Emotions, and Everyday Life*. Cambridge, MA: MIT Press.

Northcott, Michael S. 2013. *A Political Theology of Climate Change*. Grand Rapids, MI: William. B. Eerdmans.

Northcott, Michael. 2017. "On Going Gently into the Anthropocene." In *Religion in the Anthropocene*, ed. Celia Deane-Drummond, Sigurd Bergmann, Markus Vogt, and Heinrich Bedford-Strohm, 19–34. Cambridge: Lutterworth.

Norton, Bryan G. 2005. *Sustainability: A Philosophy of Adaptive Ecosystem Management*. Chicago: The University of Chicago Press.

Novonty, Patrick. 2000. "Conclusion: The Environmental Crisis Impacts Us All but Not Equally: Language, Ideology, and the Contested Narrative of the Environment in the Environmental Justice Movement." In *Where We Live Work and Play: The Environmental Justice Movement and the Struggle for a New Environmentalism*, 85–94. Westport, CT: Prager.

Nussbaum, Martha Craven. 2001. *Upheavals of Thought: The Intelligence of Emotions*. Cambridge: Cambridge University Press.

Nussbaum, Martha C. 2004. *Hiding from Humanity: Disgust, Shame, and the Law*. Princeton, NJ: Princeton University Press.

Nussbaum, Martha C. 2006. *Frontiers of Justice: Disability, Nationality, Species Membership, Tanner Lectures on Human Values*. Cambridge, MA: The Belknap Press of Harvard University Press.

Nussbaum, Martha Craven. 2013. *Political Emotions: Why Love Matters for Justice*. Cambridge, MA: The Belknap Press of Harvard University Press.

Nussbaum, Martha Craven. 2016. *Anger and Forgiveness: Resentment, Generosity, Justice*. Oxford: Oxford University Press.

O'Brien, Kevin J. 2017. *The Violence of Climate Change*. Washington, DC: Georgetown University Press.

Oates, Caroline J., and Seonaidh McDonald. 2006. "Recycling and the Domestic Division of Labour: Is Green Pink or Blue?" *Sociology* 40(3): 417–33.

Oceti Sakowin Camp. 2016. "Forgiveness Ceremony: Veterans Kneel at Standing Rock." YouTube.

Onwezen, Marleen C. 2015. "I Did Good, and We Did Bad: The Impact of Collective Versus Private Emotions on Pro-Environmental Food Consumption." *Food Research International* 76(2): 261–8.

Onwezen, Marleen C., Gerrit Antonides, and Jos Bartels. 2013. "The Norm Activation Model: An Exploration of the Functions of Anticipated Pride and Guilt in Pro-Environmental Behaviour." *Journal of Economic Psychology* 39: 141–53.

Orange, Donna M. 2017. *Climate Crisis, Psychoanalysis, and Radical Ethics*. Abingdon: Routledge.

Pattison, James. 2015. "The Ethics of Diplomatic Criticism: The Responsibility to Protect, Just War Theory and Presumptive Last Resort." *European Journal of International Relations* 21(4): 935–57.

Pearce, Fred. 2008. *Confessions of an Eco-Sinner: Tracking Down the Sources of My Stuff*. Boston, MA: Beacon Press.

Pew Research Center. 2019. "In U.S., Decline of Christianity Continues at a Rapid Pace: An Update on America's Changing Religious Landscape." Pew Research Forum. Accessed September 9, 2020. https://www.pewforum.org/2019/10/17/in-u-s-decline-of-christianity-continues-at-rapid-pace/.

Pihlström, Sami. 2011. *Transcendental Guilt: Reflections on Ethical Finitude*. New York: Lexington Books.

Pike, Sarah. 2016a. "Mourning Nature: The Work of Grief in Radical Enviromentalism." "Religion, Science and the Future," Gainesville, FL, January 14–16.

Pike, Sarah M. 2016b. "Mourning Nature: The Work of Grief in Radical Environmentalism." *Journal for the Study of Religion, Nature & Culture* 10(4): 419–41.

Pike, Sarah M. 2020. "Climate Strikes as Rites of Mourning the Future." SSRC. http://tif.ssrc.org/2020/02/24/climate-strikes-as-rites-of-mourning-the-future/.

Pinnington, Adrian. 2001. "Yoshimitsu, Benedict, Endō: Guilt, Shame and the Post-War Idea of Japan." *Japan Forum* 13(1): 91–105.

Piraro, Dan. 2012. "Bizarro." *The Funny Times*, 9. Distributed by King Features Syndicate, Inc, Cleveland Heights, Ohio.

Pleasants, Jen. 2009. *Bag Green Guilt*. Show the Love Media LLC.

Preston, Brian J. 2011. "The Use of Restorative Justice for Environmental Crime." EPA Victoria Seminar on Restorative Environmental Justice, Melbourne, Australia.

Preston, Stephanie D., and Frans B. de Waal. 2002. "Empathy: Its Ultimate and Proximate Bases." *Behavioral and Brain Sciences* 25: 1–72.

Price, Hilary B. 2015. "The Bounty." *Funny Times*, April 2.

Pruneau, Diane, André Doyon, Liette Vasseur, Joanne Langis, Eileen Ouellet, Elizabeth McLaughlin, Gaston Boudreau, and Gilles Martin. 2006. "When Teachers Adopt Environmental Behaviors in the Aim of Protecting the Climate." *Journal of Environmental Education* 37(3): 3–12.

Qi, Danyi, and Brian E. Roe. 2016. "Household Food Waste: Multivariate Regression and Principal Components Analyses of Awareness and Attitudes among U.S. Consumers." *PLoS ONE*: 1–19.

Quora. 2012. "Are 'Scientific' Activist Atheists Possibly Psychologically Projecting/Motivated in-Part by Hidden 'Global Warming Guilt'?" Quora. Accessed March 27. http://www.quora.com/Are-scientific-activist-atheists-possibly-psychologically-projecting-motivated-in-part-by-hidden-global-warming-guilt.

Rand, David, and Martin Nowak. 2009. "Name and Shame." *New Scientist* 204(2734): 28–9.

Ravitz, Jessica. 2016. "Not All the Standing Rock Sioux Are Protesting the Pipeline." CNN. Accessed February 25, 2020. http://www.cnn.com/2016/10/29/us/dakota-pipeline-standing-rock-sioux/.

Rees, Jonas H., Sabine Klug, and Sebastian Bamberg. 2015. "Guilty Conscience: Motivating Pro-environmental Behavior by Inducing Negative Moral Emotions." *Climatic Change* 130(3): 439–52.

Rensmann, Lars. 2004. "Collective Guilt, National Identity, and Political Processes in Contemporary Germany." In *Collective Guilt: International Perspectives*, ed. Nyla R. Branscombe and Bertjan Doosje, 169–90. New York: Cambridge University Press.

Revesz, Rachael. 2016. "13-Year-Old Native American's Petition to Stop Oil Pipeline Reaches 80,000 Signatures." *Independent*, May 11. Accessed February 24, 2020. https://www.independent.co.uk/news/world/americas/13-year-old-native-american-s-petition-to-stop-oil-pipeline-reaches-80000-signatures-a7024426.html.

Revkin, Andrew C. 2007. "Carbon-Neutral Is Hip, but Is It Green?" *The New York Times*, April 29.

Rijnen, Harry. 2003. "Business Travel: Offsetting Environmental Damage by Planes." *The New York Times*. Accessed October 16, 2008. http://query.nytimes.com/gst/fullpage.html?res=9C04E3D81E3AF93BA25751C0A9659C8B63&scp=2&sq=environmental%20guilt&st=cse

Rizzolatti, Giacomo. 2005. "The Mirror Neuron System and Its Function in Humans." *Anatomy and Embryology* 210(5–6): 419–21.

Roach, Catherine M. 2003. *Mother/Nature: Popular Culture and Environmental Ethics*. Bloomington: Indiana University Press.

Roark, Robyn. 2012. "San Jose's Plastic Bag Ban: Good or Bad for Business?" *Silicon Valley Mamas*. Accessed February 25, 2020. http://siliconvalleymamas.com/2012/02/san-joses-plastic-bag-ban-good-or-bad-for-business/.

Roccas, Sonia, Yechiel Klar, and Ido Liviatan. 2004. "Exonerating Cognitions, Group Identification, and Personal Values as Predictors of Collective Guilt among Jewish-Israelis." In *Collective Guilt: International Perspectives*, ed. Nyla R. Branscombe and Bertjan Doosje, 130–47. New York: Cambridge University Press.

Rockstrom, Johan, et al. 2009. "A Safe Operating Space for Humanity." *Nature* 461(24): 472–5.

Rogers, Heather. 2005. *Gone Tomorrow: The Hidden Life of Garbage*. New York: The New Press.

Rolling Stone. 2015. "See Hundreds of Thousands March for Action against Climate Change."

Ronson, Jon. 2015. *So You've Been Publicly Shamed*. New York: Riverhead Books.

Rosner, David, and Gerald Markowitz. 2008. "'A Problem of Slum Dwellings and Relatively Ignorant Parents': A History of Victim Blaming in the Lead Pigment Industry." *Environmental Justice* 1(3): 159–68.

Ruddiman, William F. 2013. "The Anthropocene." *Annual Review of Earth and Planetary Sciences* 41: 45–68.
Sammon, Alexander. 2016. "A History of Native Americans Protesting the Dakota Access Pipeline." *Mother Jones*.
Sanford, A. Whitney. 2007. "Pinned on Karma Rock: Whitewater Kayaking as Religious Experience." *Journal of the American Academy of Religion* 75(4D): 875–95.
Saraswati, L. Ayu. 2012. "'Malu': Coloring Shame and Shaming the Color of Beauty in Transnational Indonesia." *Feminist Studies* 38(1): 113–40.
Scheff, Thomas J. 2000. "Shame and the Social Bond: A Sociological Theory." *Sociological Theory* 18(1): 84–99.
Schmitt, Michael T., Nyla R. Branscombe, and Jack W. Brehm. 2004. "Gender Inequality and the Intensity of Men's Collective Guilt." In *Collective Guilt: International Perspectives*, ed. Nyla R. Branscombe and Bertjan Doosje, 75–92. New York: Cambridge University Press.
Schmitt, Michael T., Daniel A. Miller, Nyla R. Branscombe, and Jack W. Brehm. 2008. "The Difficulty of Making Reparations Affects the Intensity of Collective Guilt." *Group Processes & Intergroup Relations* 11(3): 267–79.
Schoppe-Sullivan, Sarah. 2016. "Worrying About Being the Perfect Mother Makes It Harder to Be a Good Parent." *Conversation*. Accessed February 25, 2020. https://theconversation.com/worrying-about-being-a-perfect-mother-makes-it-harder-to-be-a-good-parent-58690.
Schoppe-Sullivan, Sarah J., Jill E. Yavorsky, Mitchell K. Bartholomew, Jason M. Sullivan, Meghan A. Lee, Claire M. Kamp Dush, and Michael Glassman. 2016. "Doing Gender Online: New Mothers' Psychological Characteristics, Facebook Use, and Depressive Symptoms." *Sex Roles*: 1–14.
Schrader, Adam. 2016. "Ivan Looking Horse Speaks at a Forgiveness Ceremony at the Standing Rock Casino." YouTube.
Schrivan, Maria. 2012. "Suburban Dictionary: 'Eco Self-Conscious.'" *Funny Times*, April, 9.
Sengupta, Somini. 2019. "Protesting Climate Change, Young People Take to Streets in a Global Strike." *The New York Times*, September 20. Accessed October 4, 2019. https://www.nytimes.com/2019/09/20/climate/global-climate-strike.html.
Shelton Group. 2012. "Americans' Biggest 'Green' Guilt: Wasting Food, Survey Finds." Knoxville, TN.
Sideris, Lisa H. 2003. *Environmental Ethics, Ecological Theology, and Natural Selection*. New York: Columbia University Press.
Simpson, Aislinn. 2009. "Hundreds of Heathrow Protesters Sit Down for Cake and Champagne." *The Telegraph*. Accessed October 4, 2019. https://www.telegraph.co.uk/travel/travelnews/4225327/Hundreds-of-Heathrow-protests-sit-down-for-cake-and-champagne.html.
Slaby, Jan. 2014. "Emotions and the Extended Mind." In *Collective Emotions : Perspectives from Psychology, Philosophy, and Sociology*, ed. Christian von Scheve and Mikko Salmela, 32–46. Oxford: Oxford University Press.
Smart, Ninian. 1996. *Dimensions of the Sacred: An Anatomy of the World's Beliefs*. Berkeley: University of California Press.
Smith, Alexandra Nutter. 2010a. "The Ecofetish: Green Consumerism in Women's Magazines." *Women's Studies Quarterly* 38(3): 66–83.
Smith, Jonathan Z. 1980. "The Bare Facts of Ritual." *History of Religions* 20(1/2): 112–27.
Smith, Jonathan Z. 1987. *To Take Place: Toward Theory in Ritual*. Chicago: The University of Chicago Press.

Smith, Kevin. 2007. *Carbon Neutral Myth: Offset Indulgences for Your Climate Sins*. Amsterdam: Transnational Institute, Carbon Trade Watch.

Smith, Tovia. 2010b. "Companies 'Named and Shamed' for Bad Behavior." *NPR*, Weekend Edition Sunday.

Somma, Mark, and Sue Tolleson-Rinehart. 1997. "Tracking the Elusive Green Women: Sex, Environmentalism, and Feminism in the United States and Europe." *Political Research Quarterly* 50(1): 153–69.

Sonenshein, Scott, Katherine A. DeCelles, and Jane E. Dutton. 2014. "It's Not Easy Being Green: The Role of Self-Evaluations in Explaining Support of Environmental Issues." *Academy of Management Journal* 57(1): 7–37.

Standing Rock Sioux Tribe. 2017. "Camp Information: Etiquette." Stand with Standing Rock. http://standwithstandingrock.net/camp-etiquette/.

Standing Rock Veterans. 2016. "The Wes Clark Jr. Show." *Medium*, December 10.

Statista. 2017. *Largest Producers of CO2 Emissions Worldwide in 2016, Based on Their Share of Global CO2 Emissions*. Germanwatch.

Steffen, Will, et al. 2011a. "The Anthropocene: From Global Change to Planetary Stewardship." *Ambio* 40(7): 739–61.

Steffen, Will, et al. 2011b. "The Anthropocene: From Global Change to Planetary Stewardship." *Ambio* (7): 739.

Stephan, Mark. 2002. "Environmental Information Disclosure Programs: They Work, but Why?" *Social Science Quarterly* 83(1): 190–205.

Stewart, Rachel. 2016. "Emerging from the Mire: Using Public Shaming after an Oil Spill to Yield Informed Decisions and Alter Behavior." *Vermont Law Review* 40(3): 695–732.

Stipek, Deborah. 1995. "The Development of Pride and Shame in Toddlers." In *Self-Conscious Emotions: The Psychology of Shame, Guilt, Embarrassment, and Pride*, ed. June Price Tangney and Kurt W. Fischer, 237–52. New York: The Guilford Press.

Streeter, A. K. 2008. "Eco-Guilt the New Plague among Enlightened Parents." Discovery Communications, LLC. Accessed February 25, 2020. http://www.treehugger.com/green-food/eco-guilt-the-new-plague-among-enlightened-parents.html.

Strickland, Cecily Y., and Scott (Rusty) Miller. 2007. "Creative Sentencing, Restorative Justice and Environmental Law: Responding to the Terra Nova FPSO Oil Spill." *Dalhousie Law Journal* 30(2): 547ff.

Sullivan, Gavin Brent. 2015. "Collective Emotions." *Social and Personality Psychology Compass* 9(8): 383–93.

Sullivan, Shannon. 2014. *Good White People : The Problem with Middle-Class White Anti-Racism*. Albany, NY: State University of New York Press.

Suplico, Luz T. 2009. "Impact of Green Marketing on the Students' Purchase Decision." *Journal of International Business Research* 8(SI.2): 71.

Taliman, Valerie. 2016a. "Thousands of Veterans Descend on Standing Rock to Protect and Serve." *Indian Country Today*. Accessed September 16, 2020. https://indiancountrytoday.com/archive/thousands-of-veterans-descend-on-standing-rock-to-protect-and-serve-CsmzQI5XsEOz9ZyQf2WoZw

Taliman, Valerie. 2016b. "Veterans Ask for Forgiveness and Healing in Standing Rock." *Indian Country Today*. Accessed September 16, 2020. https://indiancountrytoday.com/archive/veterans-ask-for-forgiveness-and-healing-in-standing-rock-k2fLUonhqEy6D-NjnhlJJw

Tangney, June Price. 1995. "Shame and Guilt in Interpersonal Relationships." In *Self-Conscious Emotions: The Psychology of Shame, Guilt, Embarrassment, and Pride*, ed. June Price Tangney and Kurt W. Fischer, 114–39. New York: The Guilford Press.

Tangney, June Price, Susan A. Burggraf, and Patricia E. Wagner. 1995. "Shame-Proneness, Guilt-Proneness, and Psychological Symptoms." In *Self-Conscious Emotions: The Psychology of Shame, Guilt, Embarrassment, and Pride*, ed. June Price Tangney and Kurt W. Fischer, 343–67. New York: The Guilford Press.

Tangney, June Price, and Ronda L. Dearing. 2002. *Shame and Guilt*. New York: The Guilford Press.

Tangney, June Price, and Kurt W. Fischer, eds. 1995. *Self-Conscious Emotions: The Psychology of Shame, Guilt, Embarrassment, and Pride*. New York: The Guilford Press.

Tangney, June Price, Jeffrey Stuewig, and Debra J. Mashek. 2007. "What's Moral About the Self-Conscious Emotions?" In *The Self-Conscious Emotions: Theory and Research*, ed. Jessica L. Tracy, Richard W. Robins, and June Price Tangney, 21–37. New York: The Guilford Press.

Tass. 2008. "What Gives You the Most 'Green' Guilt?". Babyworld Forums. Accessed February 25, 2020. http://babyworld.co.uk/groups/eco-parenting/forum/topic/379662-what-gives-you-the-most-green-guilt/.

Taylor, Bron. 2007a. "Exploring Religion, Nature and Culture: Introducing the *Journal for the Study of Religion, Nature and Culture*." *Journal for the Study of Religion, Nature and Culture* 1(1): 5–24.

Taylor, Bron. 2007b. "Focus Introduction: Aquatic Nature Religion." *Journal of the American Academy of Religion* 75(4D): 863–74.

Taylor, Bron. 2010. *Dark Green Religion: Nature Spirituality and the Planetary Future*. Berkeley: University of California Press.

Taylor, Erin N., and Lora Ebert Wallace. 2012. "For Shame: Feminism, Breastfeeding Advocacy, and Maternal Guilt." *Hypatia* 27(1): 76.

Taylor, Matthew, Jonathan Watts, and John Barlett. 2019. "Climate Crisis: 6 Million People Join Latest Wave of Global Protests." *The Guardian*, September 27. Accessed September 27, 2019. https://www.theguardian.com/environment/2019/sep/27/climate-crisis-6-million-people-join-latest-wave-of-worldwide-protests.

TerraChoice Environmental Marketing. 2007. "The Six Sins of Greenwashing: A Study of Environmental Claims in North American Consumer Markets."

Tessman, Lisa. 2005. *Burdened Virtues: Virtue Ethics for Liberatory Struggles*. Oxford: Oxford University Press.

Thomas, Tandy Chalmers, Rebecca K. Trump, and Linda L. Price. 2015. "Advertising as Unfavorable Self-Preservation: The Dirty Laundry Effect." *Journal of Advertising* 44(1): 58–70.

Thomason, Krista K. 2018. *Naked : The Dark Side of Shame and Moral Life*. New York: Oxford University Press.

Thornbecke, Catherine. 2016. "Timeline of Events Surrounding the Dakota Access Pipeline." CBS. http://abcnews.go.com/US/timeline-dakota-access-pipeline-protests/story?id=43131355.

Thunberg, Greta. 2019. "'You Did Not Act in Time'" Greta Thunberg's Full Speech to MPs." *The Guardian*, April 23. Accessed February 25, 2020. https://www.theguardian.com/environment/2019/apr/23/greta-thunberg-full-speech-to-mps-you-did-not-act-in-time

Tian, Xian-Liang, Qi-Guang Guo, Chao Han, and Najid Ahmad. 2016. "Different Extent of Environmental Information Disclosure across Chinese Cities: Contributing Factors and Correlation with Local Pollution." *Global Environmental Change* 39: 244–57.

Tilikidou, Irene, and Antonia Delistavrou. 2008. "Types and Influential Factors of Consumers' Non-purchasing Ecological Behaviors." *Business Strategy and the Environment* 17(1): 61–76.

Tim. 2010. "Green Survey Provides Cause for New Year Optimism." *Green Gazette*.

Tjernshaugen, Andreas, and Ho-Ching Lee. 2007. "Shaming and Framing: Norwegian NGOs in the Climate Change Negotiations." In *Europe and Global Climate Change: Politics, Foreign Policy and Regional Cooperation*, ed. Paul G. Harris, 185–210. Northampton, MA: Edward Elgar.

Tolan, Sandy. 2016. "Veterans Came to North Dakota to Protest a Pipeline. But They Also Found Healing and Forgiveness." *Los Angeles Times*.

Trump, Donald J. 2017. "Presidential Memorandum Regarding Construction of the Dakota Access Pipeline." Washington, DC, January 24.

Tucker, Elizabeth M., Nora J. Rifon, Eun Mi Lee, and Bonnie B Reece. 2012. "Consumer Receptivity to Green Ads: A Test of Green Claim Types and the Role of Individual Consumer Characteristics for Green Ad Response." *Journal of Advertising* 41(4): 9–23.

Turner, James. 2013. "Op Ed: The Climate Change Guilt Trip: The Battle against Global Warming Should Not Be About Judging People's Every Choice." *Los Angeles Times*. http://articles.latimes.com/2013/jun/02/opinion/la-oe-turner-climate-change-20130602.

Turner, John C. 1982. "Towards a Cognitive Redefinition of the Social Group." In *Social Identity and Intergroup Relations*, ed. Henri Tajfel, 15–40. Cambridge: Cambridge University Press.

Unicorn Riot. 2016. "Veterans Apologize for Genocide & March to Backwater Bridge in Blizzard." *Unicorn Riot*.

University of Michigan Institute for Social Research. 2008. "Chore Wars: Men, Women and Housework." National Science Foundation.

van Erp, Judith. 2014. "Naming and Shaming of Corporate Offenders." In *Encyclopedia of Criminology and Criminal Justice*, ed. Gerben Bruinsma and David Weisburd, 3209–17. New York: Springer.

van Gelder, Sarah. 2016. "Why I Kneeled before Standing Rock Elders and Asked for Forgiveness." *YES! Magazine*.

Van Vugt, Mark. 2009. "Averting the Tragedy of the Commons: Using Social Psychological Science to Protect the Environment." *Current Directions in Psychological Science* 18(3): 169–73.

van Wieren, Gretel. 2013. *Restored to Earth: Christianity, Environmental Ethics, and Ecological Restoration*. Washington, DC: Georgetown University Press.

Vannier-Santos, Marcos A., and Henrique L. Lenzi. 2011. "Parasites or Cohabitants: Cruel Omnipresent Usurpers or Creative 'Éminences Grises'?" *Journal of Parasitology Research* 2011: 214174.

Vaze, Prashant. 2009. *The Economical Environmentalist: My Attempt to Live a Low-Carbon Life and What It Cost*. London: Earthscan.

Velasquez, Manuel. 1991. "Why Corporations Are Not Morally Responsible for Anything They Do." In *Collective Responsibility: Five Decades of Debate in Theoretical and Applied Ethics*, ed. Larry May and Stacey Hoffman, 111–31. Lanham, MD: Rowman & Littlefield.

Verdier, Catherine. 2007. "Carbon Offsets: Are Such Credits Effectively Helping Mitigate Climate Change?" *Sustainable Development Law & Policy* 7(2): 63.

Verry, John, Felicity Heffernan, and Richard Fisher. 2005. "Restorative Justice Approaches in the Context of Environmental Prosecution." ABS House, Canberra, June 6–7.

Veterans Stand. 2017. "Veterans Stand: Transform Tour '17." Accessed August 4, 2017. https://veterans-stand.org/mission/.

Vinz, Dagmar. 2009. "Gender and Sustainable Consumption: A German Environmental Perspective." *European Journal of Women's Studies* 16: 159–79.

von Jessen, Fritzie. 2009. *I Killed a Penguin: An Ecological Memoir*. Baltimore, MD: Publish America.

von Scheve, Christian, and Sven Ismer. 2013. "Towards a Theory of Collective Emotions." *Emotion Review* 5(4): 406–13.

von Scheve, Christian, and Mikko Salmela, eds. 2014. *Collective Emotions : Perspectives from Psychology, Philosophy, and Sociology*. Oxford: Oxford University Press.

Wadler, Joyce. 2010. "Green, But Still Feeling Guilty." *The New York Times*, September 29. Accessed April 16, 2012. http://www.nytimes.com/2010/09/30/garden/30guilt.html?_r=1.

Wallbott, Harald G., and Klaus R. Scherer. 1995. "Cultural Determinants in Experiencing Shame and Guilt." In *Self-Conscious Emotions: The Psychology of Shame, Guilt, Embarrassment, and Pride*, ed. June Price Tangney and Kurt W. Fischer, 465–87. New York: The Guilford Press.

Wang, Zhaohua, and Chen Wang. 2014. "How Carbon Offsetting Scheme Impacts the Duopoly Output in Production and Abatement: Analysis in the Context of Carbon Cap-and-Trade." *Journal of Cleaner Production* 103: 715–23.

Warren, Roz. 2014. "Greener Than Thou: Saving the Earth One Smackdown at a Time." *The Funny Times*, April, 4.

Waters, Colin N., Jan A. Zalasiewicz, Mark Williams, Michael A. Ellis, and Andrea M. Snelling. 2014. "A Stratigraphical Basis for the Anthropocene?" In *A Stratigraphical Basis for the Anthropocene*, ed. Colin N. Waters, Jan A. Zalasiewicz, Mark Williams, Michael A. Ellis and Andrea M. Snelling, 1–21. London: Geological Society of London.

wcmctv. 2016. "Wesley Clark Jr Asks for Forgiveness from Leonard Crow Dog." YouTube: YouTube.

Wehelie, Benazir. 2016. "Sacred Ground: Inside the Dakota Pipeline Protests." Accessed February 24, 2020. https://www.cnn.com/interactive/2016/12/us/dapl-protests-cnnphotos/.

Weinersmith, Kelly L., and Ryan L. Earley. 2016. "Better with Your Parasites? Lessons for Behavioural Ecology from Evolved Dependence and Conditionally Helpful Parasites." *Animal Behaviour* 118: 123–33.

Weintrobe, Sally, ed. 2013. *Engaging with Climate Change: Psychoanalytic and Interdisciplinary Perspectives, New Library of Psychoanalysis*. New York: Routledge.

Weisman, Alan. 2007. *The World Without Us*. New York: Thomas Dunne Books/St. Martin's Press.

Weisman, Tama. 2012. "Restorative Environmental Justice as a Way of Life: Learning from Ubuntu." *Dialogue and Universalism* 3(1): 92–109.

West, Traci C. 2006. *Disruptive Christian Ethics: When Racism and Women's Lives Matter*. Louisville, KY: Westminster John Knox Press.

Whyte, Kyle Powys. 2017. "Is It Colonial Deja Vu? Indigenous Peoples and Climate Justice." In *Humanities for the Environment: Integrating Knowledge, Forging New Constellations of Practice*, ed. Joni Adamson and Michael Davis, 88–105. New York: Routledge.

Williams, Alex, with contributions by Catherine Donaldson-Evans, Amy Goetzman, Kate Hammer, Carol Pogash, Rachel Pomerance, and Paula Schwartz. 2007. "Water, Water Everywhere, but Guilt by the Bottleful." *The New York Times*. Accessed October 16, 2008. http://www.nytimes.com/2007/08/12/fashion/12water.html?scp=1&sq=environmental%20guilt&st=cse

Willis, Jennifer. 2008. "Goodbye Green Guilt, Hello Change: Next Year Will Be All About Making Small Changes to Help the Planet." *The Home Forum*, April 22. http://www.csmonitor.com/The-Culture/The-Home-Forum/2008/1229/p19s03-hfes.html.

Wilson, Edward O. 1993a. "Is Humanity Suicidal?" *The New York Times Magazine*, May 30, 24–8.

Wilson, Edward O. 1993b. "Is Humanity Suicidal?" *BioSystems* 31: 235–42.

Wohl, Michael J. A., Kimberly Matheson, Nyla R. Branscombe, and Hymie Anisman. 2013. "Victim and Perpetrator Groups' Responses to the Canadian Government's Apology for the Head Tax on Chinese Immigrants and the Moderating Influence of Collective Guilt." *Political Psychology* 34(5): 713–29.

Wolf, Paige. 2010. *Spit That Out: The Overly Informed Parent's Guide to Raising Children in the Age of Environmental Guilt*. Charleston, S.C: Lombard Books.

Wolk-Stanley, Katy. 2012a. "A Questioning Day." *The Non-Consumer Advocate*, October 18. Accessed February 25, 2020. https://thenonconsumeradvocate.com/a-questioning-day/

Wolk-Stanley, Katy. 2012b. "A Rodent Induced Non-consumer Confession." *The Non-Consumer Advocate*. Accessed February 25, 2020. https://thenonconsumeradvocate.com/a-rodent-induced-non-consumer-confession/

Wong, Julia Carrie. 2016. "Dakota Access Pipeline: US Denies Key Permit, a Win for Standing Rock Protesters." *The Guardian*, December 5. https://www.theguardian.com/us-news/2016/dec/04/dakota-access-pipeline-permit-denied-standing-rock.

Wong, Ying, and Jeanne Tsai. 2007. "Cultural Models of Shame and Guilt." In *The Self-Conscious Emotions: Theory and Research*, ed. Jessica L. Tracy, Richard W. Robins, and June Price Tangney, 209–23. New York: The Guilford Press.

Yang, Kenneth C. C., Marissa Jiménez, and Kang Yowei. 2015. "The Influence of Ethnicity on Consumers' Environmental Attitudes and Responses to Guilt Appeals in Green Advertising: An Experimental Study." *Intercultural Communication Studies* 24(2): 192–206.

Yates, Adam, Ye Luo, Catherine Mobley, and Erin Shealy. 2015. "Changes in Public and Private Environmentally Responsible Behaviors by Gender: Findings from the 1994 and 2010 General Social Survey." *Sociological Inquiry* 85(4): 503–31.

Yates, Emily. 2017. "We Are Not at War." 0-Dark-Thirty. Accessed August 5. https://o-dark-thirty.org/2017/08/01/we-are-not-at-war/.

Yoneyama, Lisa. 1999. "Habits of Knowing Cultural Differences: Chrysanthemum and the Sword in the U.S. Liberal Multiculturalism." *Topoi* 18(1): 71–80.

Young, Wendy. 2012. "Living Green—Guilt Free!" Accessed April 16. http://www.simple-green-living.com/living-green.html.

Zahn-Waxler, CArolyn, and JoAnn Robinson. 1995. "Empathy and Guilt: Early Origins of Feelings of Responsibility." In *Self-Conscious Emotions: The Psychology of Shame, Guilt, Embarrassment, and Pride*, ed. June Price Tangney and Kurt W. Fischer, 143–73. New York: The Guilford Press.

Zalasiewicz, Jan, et al. 2015. "When Did the Anthropocene Begin? A Mid-Twentieth Century Boundary Level Is Stratigraphically Optimal." *Quaternary International* 383: 196–203.

Zebel, Sven, Bertjan Doosje, and Russell Spears. 2004. "It Depends on Your Point of View: Implications of Perspective-Taking and National Identification for Dutch Collective Guilt." In *Collective Guilt: International Perspectives*, ed. Nyla R. Branscombe and Bertjan Doosje, 148–68. New York: Cambridge University Press.

Zeller Jr., Tom. 2009. "Thousands March in Copenhagen, Calling for Action." *The New York Times*. Accessed October 4, 2019. https://www.nytimes.com/2009/12/13/science/earth/13climate.html.

Zhou, Naaman. 2018. "Climate Change Strike: Thousands of School Students Protest across Australia." *The Guardian*, November 30. Accessed October 4, 2019. https://www.theguardian.com/environment/2018/nov/30/climate-change-strike-thousands-of-students-to-join-national-protest.

Zylicz, Piotr Olaf, and Ludmila Poleszak. 2005. "Naturalistic Conception of Collective Guilt " *Humboldt Journal of Social Relations* 29(2): 185–206.

Index

Note: Tables are indicated by an italic "*t*" following the page number.

For the benefit of digital users, indexed terms that span two pages (e.g., 52–53) may, on occasion, appear on only one of those pages.